MODERN VULNERABILITY MANAGEMENT

Predictive Cybersecurity

For a complete listing of titles in the
Artech House Computer Security Library,
turn to the back of this book.

MODERN VULNERABILITY MANAGEMENT

Predictive Cybersecurity

Michael Roytman
Ed Bellis

ARTECH HOUSE
BOSTON | LONDON
artechhouse.com

Library of Congress Cataloging-in-Publication Data
A catalog record for this book is available from the U.S. Library of Congress.

British Library Cataloguing in Publication Data
A catalogue record for this book is available from the British Library.

Cover design by Andy Meaden

ISBN 13: 978-1-63081-938-5

© 2023
Artech House
685 Canton Street
Norwood, MA 02062

All rights reserved. Printed and bound in the United States of America. No part of this book may be reproduced or utilized in any form or by any means, electronic or mechanical, including photocopying, recording, or by any information storage and retrieval system, without permission in writing from the publisher.

All terms mentioned in this book that are known to be trademarks or service marks have been appropriately capitalized. Artech House cannot attest to the accuracy of this information. Use of a term in this book should not be regarded as affecting the validity of any trademark or service mark.

10 9 8 7 6 5 4 3 2 1

CONTENTS

FOREWORD *XI*

ACKNOWLEDGMENTS *XV*

1
THE STATE OF THE VULNERABILITY LANDSCAPE 1

1.1	The Security Canon: Fundamental Cybersecurity Terminology	4
	1.1.1 Common Vulnerabilities and Exposures	5
	1.1.2 National Vulnerability Database	7
	1.1.3 Common Vulnerability Scoring System	7
	1.1.4 Common Weakness Enumeration	7
	1.1.5 Common Platform Enumeration	7
1.2	Security Metrics: The New Guard	8
	References	13

2
DATA SCIENCE TO DEFINE RISK 15

2.1	Risk Management History and Challenges	15
	2.1.1 The Birth of Operations Research	16
	2.1.2 The Scale of Cybersecurity	18

2.1.3 Origins of the Risk-Based Approach to Vulnerability Management 20
References 24

3
DECISION SUPPORT: TAPPING MATHEMATICAL MODELS AND MACHINE LEARNING 25

3.1 Mathematical Modeling 26
    3.1.1 Mathematical Scale 27
    3.1.2 Statistics 29
    3.1.3 Game Theory 32
    3.1.4 Stochastic Processes 34
    3.1.5 OODA Loops 37
3.2 Machine Learning for Cybersecurity 38
    3.2.1 Supervised Models 39
    3.2.2 Unsupervised Models 40
    References 45

4
HOW TO BUILD A DECISION ENGINE TO FORECAST RISK 47

4.1 The Data 48
    4.1.1 Definitions vs Instances 50
    4.1.2 Vulnerability Data 55
    4.1.3 Threat Intel Sources 60
    4.1.4 Asset Discovery and Categorization: Configuration Management Database 62
    4.1.5 Data Validation 64
4.2 Building a Logistic Regression Model 65
    4.2.1 Data Sources and Feature Engineering 66
    4.2.2 Testing Model Performance 69
    4.2.3 Implementing in Production 72
4.3 Designing a Neural Network 79
    4.3.1 Preparing the Data 79

4.3.2	Developing a Neural Network Model	82
4.3.3	Hyperparameter Exploration and Evaluation	84
4.3.4	Scoring	95
4.3.5	Future Work	100
References		101

5

MEASURING PERFORMANCE 103

5.1	Risk vs Performance	104
5.2	What Makes a Metric "Good"?	105
	5.2.1 Seven Characteristics of Good Metrics	106
	5.2.2 Evaluating Metrics Using the Seven Criteria	108
	5.2.3 More Considerations for Good Metrics	110
5.3	Remediation Metrics	111
	5.3.1 Mean-Time-Tos	111
	5.3.2 Remediation Volume and Velocity	112
	5.3.3 R Values and Average Remediation Rates	114
5.4	Why Does Performance Matter?	118
5.5	Measuring What Matters	119
	5.5.1 Coverage and Efficiency	119
	5.5.2 Velocity and Capacity	123
	5.5.3 Vulnerability Debt	132
	5.5.4 Remediation SLAs	135
References		139

6

BUILDING A SYSTEM FOR SCALE 141

6.1	Considerations Before You Build	141
	6.1.1 Asset Management Assessment	143
	6.1.2 Where Your Organization Is Going	144
	6.1.3 Other Tools as Constraints	145
6.2	On Premise vs Cloud	146
6.3	Processing Considerations	147
	6.3.1 Speed of Decisions and Alerts	147

	6.3.2	SOC Volume	149
6.4	Database Architecture		150
	6.4.1	Assets Change Faster Than Decisions	151
	6.4.2	Real-Time Risk Measurement	152
6.5	Search Capabilities		154
	6.5.1	Who Is Searching?	154
6.6	Role-Based Access Controls		156

7
ALIGNING INTERNAL PROCESS AND TEAMS   159

7.1	The Shift to a Risk-Based Approach		160
	7.1.1	Common Goals and Key Risk Measurements	160
	7.1.2	Case Study: More Granular Risk Scores for Better Prioritization	162
7.2	Driving Down Risk		164
	7.2.1	Aligning Teams with Your Goals	165
	7.2.2	The Importance of Executive Buy-In	166
	7.2.3	Reporting New Metrics	167
	7.2.4	Gamification	167
7.3	SLA Adherence		168
	7.3.1	High-Risk vs Low-Risk Vulnerabilities	169
	7.3.2	When to Implement or Revise SLAs	170
	7.3.3	What to Include in Your SLA	172
7.4	Shifting from Security-Centric to IT Self-Service		173
	7.4.1	How to Approach Change Management	174
	7.4.2	Enabling Distributed Decision-Making	175
	7.4.3	Signs of Self-Service Maturity	177
7.5	Steady-State Workflow		177
	7.5.1	The Limits of Remediation Capacity	177
	7.5.2	Media-Boosted Vulnerabilities	178
	7.5.3	Exception Handling	179
7.6	The Importance of Process and Teams		179

8
REAL-WORLD EXAMPLES 181

8.1 A Word from the Real World — 181
    8.1.1 Vulnerability Discovery — 182
    8.1.2 Vulnerability Assessment and Prioritization — 182
    8.1.3 Vulnerability Communication — 183
    8.1.4 Vulnerability Remediation — 184
    8.1.5 What Success Looks Like — 184

9
THE FUTURE OF MODERN VM 187

9.1 Steps Toward a Predictive Response to Risk — 188
    9.1.1 Passive Data Collection — 190
9.2 Forecasting Vulnerability Exploitation with the Exploit Prediction Scoring System — 191
9.3 Support from Intelligent Awareness — 194
9.4 The Rise of XDR — 196
9.5 The Other Side of the Coin: Remediation — 198
9.6 The Wicked Problem of Security Advances — 200
    References — 201

GLOSSARY 203

ABOUT THE AUTHORS 207

INDEX 209

FOREWORD

I became closely involved with the field of vulnerability management when I was a Gartner analyst starting in 2011. Immediately a few things became obvious to me, as I started speaking with clients and industry connections, as well as with vendors involved in the entire vulnerability management process.

First, this is an area where many practices and processes seemed to have originated in the 1990s and are modernizing very slowly, even today.

Second, this is connected to the fact that vulnerabilities, their exploitation, mitigation, and remediation—such as patching—involve a complex web of security issues as well as operational and business issues.

Third, this area struck me as the best example where security teams and other IT teams have frictions, arguments, and sometimes outright fights. I've long used this domain as the best example of fears, arguments, and battles between the silos within IT—and outside as well. (Security: "We need to fix it." IT: "This is too hard." Business: "Just accept the risk.")

Fourth, this is a fascinating area where infrastructure and application issues also come into contact, and again, often arguments and frictions as well. It is very clear that the most popular way to remediate a vulnerability in a commercial operating system is to patch it. But what about an application you wrote? What about an application that somebody wrote for you? Looking for a patch is quite difficult if the original developer of the application is no longer around and nobody knows how to fix the code.

Over the years, vulnerability management gained an unhealthy reputation among many security professionals. Those who are working for traditional, especially large, organizations are used to the scare of a 10,000-page report produced by a vulnerability scanner. These reports still occupy a painful central spot in the collective consciousness of many security professionals involved with vulnerability management.

In fact, the idea that there is a very, very, very long report produced by scanner that now needs to be transferred to the team that can actually fix vulnerabilities first arose in popular consciousness in the early 2000s. Now, in the early 2020s it still persists. This means that in some areas of vulnerability management we are looking at literally 20 years of stagnation.

One critical theme that emerged as we progressed with vulnerability management research in my analyst years was the theme of prioritization. We can debate the value and merits of a 10,000-page report with scanner findings. But what is not debatable is that no organization today can fix all vulnerabilities. Or, frankly, even find them! Some organizations naively assume that they can only fix high-severity or high-priority vulnerabilities, however defined (CVSS, of course, comes to mind here). This leads to both working too hard and not fixing the real risks sometimes, doing too much and not enough at the same time.

Looking at the vulnerability management landscape, with that in mind, it becomes obvious that the only magic in vulnerability management is in vulnerability prioritization. How do we prioritize vulnerabilities that represent real risk for your organization? Many technology vendors have tried to answer it using different methods; some looked at asset values, some tried to find exploitation paths. As a result, one effective consensus approach has not emerged yet, despite years of trying. It is very likely that data science and advanced analytics would have to be part of the answer. However, until we, as an industry, arrive at this approach, we will continue to be plagued by vulnerabilities and their consequences—as well as conflicts and burnout.

I first met Michael when I was an analyst at Gartner; I quickly realized that he is one of the very, very few people (at least at the time) with a genuine expertise in both security and data science. His work at Kenna Security was groundbreaking in terms of both value for clients and novelty of approach. Moreover, it actually worked in the real world—unlike many others' attempts to marry security and data science.

So, what is in store for vulnerability management? Frankly, many of the 1990s problems that plague organizations are expected to persist as many practices still haven't changed. I have a suspicion that when you, dear reader, refer to this book in 2025 you will recognize many familiar practices and unfortunately many familiar problems that still exist in your IT organization.

For sure, there are bright spots. Digital natives run immutable infrastructure where assets are never patched, but instead are destroyed and replaced with safer versions. Modern IT practices such as DevOps occasionally—and this word is important here—lead to rapid deployment of system changes or even changed systems (such as when immutable infrastructure is used).

Mitigation has also risen in popularity and more organizations understand that the real goal is to reduce the risk rather than merely patch the hole.

Later, a bright future is probably coming. While many organizations who use the public cloud follow traditional IT practices, and migrate virtual machines to the cloud provider networks, thus suffering the old problems, digital natives and cloud natives use technology and run their IT differently. Extrapolating forward from my analyst years, I expect that the only way to "fix" vulnerability management is to change and evolve how IT is done.

Thus, this book is exciting for me as Michael reveals impressive data science magic that works to solve the tough vulnerability management challenges described here.

Dr. Anton Chuvakin
Office of the CISO at Google Cloud
Previously VP of Research at Gartner
responsible for vulnerability management
March 2023

ACKNOWLEDGMENTS

Risk-based vulnerability management did not exist in 2010, and this book is the result of hundreds of colleagues and thousands of practitioners coming together to create a new field. This book is an idea, that became a startup, that became a product, that became a market, that has become a book. This is perhaps a roundabout way of getting there; but have no doubt as you read this book that the ideas contained have been field tested at real enterprises spanning the globe, that code exists that builds all of the models you will be exposed to, and even more code and ideas have been thrown away along the way.

A few individuals are worth highlighting first due to their outsized impact on this story and without whom this book would not be possible. We want to thank Jeff Heuer for helping pioneer this data-driven approach, applying his skills in almost every discipline and group. Karim Toubba for understanding the risk-based vulnerability market before it really existed and for continuously leading the effort with integrity and a respect for the science. Jay Jacobs, Wade Baker, and Ben Edwards made a lot of the analysis you will read in this book possible with their pioneering work at the Verizon DBIR and then subsequently on the "Prioritization to Prediction" reports at Cyentia. They brought the science to the data of vulnerability management. Data science doesn't work without experts in the loop. Without Jerry Gamblin's and Jonathan Cran's input a lot of our research process would have been the data leading the blind. The best models we built were often wrong for reasons only Jerry and Jonathan could see, and we are thankful for their vigilant eyes.

There are also over three hundred individuals who worked at Kenna Security during the development of this methodology. Each one of them, from engineering leaders to product marketing managers to sales development representatives to human resources staff have made this novel approach possible. These ideas required data at scale to test, required enterprises to trust and see results from the process and continue to optimize their practices, and none of this could be collected as thoughtfully anywhere else. We also want to acknowledge the thousands of individuals working to secure hundreds of enterprises that took a leap on a new methodology because they dove into and understood the science of risk-based vulnerability management. Without all of these people, you would be reading about an idea. Instead, you are reading about a tried and tested way to secure the world's enterprises.

This methodology was developed through feedback and suggestions provided by the countless individuals in pursuit of the intersection of risk management and information security. There are few communities so full of purpose and with such a wide range of representation, as the Society of Information Risk Analysts, the MetriCon conferences, the Conference on Applied Machine Learning in Information Security (CAMLIS), and Bsides (Chicago, San Francisco, Las Vegas). The feedback on our talks, the countless lunch, dinner, and late-night conversations, and the exchange of ideas that took place in those hallways tested and refined the approach in the best of ways.

Dan Geer's seminal measuring security slide deck foretold and blueprinted much of the work that has happened here. He uniquely foresaw the application of epidemiologic methods to most cybersecurity domains, and this realization formalized novel methods of analysts. On more than one occasion, we simply found the data necessary to quantify ideas Dr. Geer put forth a decade earlier. Early in the process, he found our data interesting and coauthored two publications based on the data. The rigor he brought to that analysis showed us what "good" looked like in this field, and we remain grateful for the opportunity to build on these ideas.

Lastly, we want to thank our families for supporting us through many uncertain years. Working in an unexplored domain comes with elevated risk. As this book will show, there are many ways to address a risk. We have been incredibly fortunate to have the support and encouragement necessary to mitigate ours along the way.

Thank you.

1

THE STATE OF THE VULNERABILITY LANDSCAPE

Late on the night of July 29, 2017, the Equifax countermeasures team updated Secure Sockets Layer (SSL) certificates in the company's Alpharetta, Georgia data center [1]. It was a task long overdue. One of those certificates, installed on the SSL Visibility (SSLV) tool monitoring Equifax's customer dispute portal, had expired 19 months earlier.

After the installation, the countermeasures team confirmed they had regained visibility for the first time in more than a year and a half. One of the first packet requests they detected was from an Internet Protocol (IP) address originating in China, one of many recent requests. The server responses to many of these requests contained more than 10 MB of data.

The next day, at 12:41 pm, Equifax shut down the consumer dispute portal after the forensic team discovered the exfiltrated data likely contained personally identifiable information (PII). In fact, the attackers had made off with the financial data of 148 million Americans, the result of an attack that went undetected and uninterrupted for 76 days.

Much has been written about the Equifax breach, arguably the most consequential data breach of all time. As with any cybersecurity incident, the postmortems arrived in spades and continued in the months and years after the attack. These postmortems certainly might help Equifax avoid a similar fate in the future. But what Equifax—and every other company—actually needs is a way to build forecasts to prevent attacks before they happen.

Why did the breach take 76 days to detect? Equifax had a mature, well-oiled security team—and it turns out they did their jobs well. They scanned their systems, they inventoried their assets, they applied patches and remediated vulnerabilities—thousands of them per month. So why not this one?

Coverage of the breach focused on how Equifax had failed to patch the Apache Struts vulnerability, CVE-2017-5638, which was publicly disclosed on March 7, 2017. That's more than two months before attackers exploited that vulnerability to gain access to Equifax's data.

We know now that the breach was a legitimate hack, but Equifax still should have seen it coming. The real reason for the miss had nothing to do with Equifax or its security team, and everything to do with the traditional approach to vulnerability management—a numbers game born in the late 1990s and early 2000s that has remained largely unquestioned since then. At the time of the Equifax attack, looking at data from more than 300 enterprises with a total of 1 billion vulnerabilities, more than a quarter of them (259,451,953) had a Common Vulnerability Scoring System (CVSS) score of 9 or 10.

So the CVSS score alone didn't make the Apache Struts vulnerability stand out. But there were other clues to its criticality. First, exploits were observed in the wild and revealed just a day after the vulnerability was made public [2]. Then, a Metasploit module was released for this vulnerability on March 15, 2017, just eight days after it was disclosed. For years, data has indicated that a vulnerability having a weaponized, public exploit available is one of the single biggest factors in predicting successful exploitations.

There were other common-sense attributes of the Apache Struts vulnerability that increased the likelihood of exploitation, including:

- The target of opportunity represented by Apache Struts;
- The breadth of affected operating systems;
- The high impact of the vulnerability;
- The potential for remote code execution.

Clearly, this wasn't just any other vulnerability. It stood out even among the quarter of vulnerabilities with the highest CVSS scores. But without a system to prioritize which vulnerabilities to fix, and to indicate exactly how critical this vulnerability was, Equifax was a sitting duck.

Usually, failures of prioritization are swept under the rug. If you fail to prioritize important tasks in your daily life, to the untrained eye it looks like nothing is happening. Sure, you could be more effective. You could have accomplished more. But nothing catastrophic occurs. When it comes to cybersecurity, and vulnerability management in particular, the scale doesn't allow for inefficiency. Enterprises have threat actors—malicious ones—and they aren't standing still. What's worse, without data-driven prioritization at scale, not only do the number of potential risks like the Apache Struts vulnerability increase over time, but so does the probability that something catastrophic does happen. To paraphrase the author Chuck Palahniuk, on a long enough timeline, every risk becomes an event.

The real story here is that every enterprise is currently at risk of the same system failure as Equifax no matter how good the security or IT teams are at their day jobs. It's the system itself that creates the potential for failure. Vulnerabilities are relentless, growing by the tens of thousands every year. The growth rate itself is growing. The process of identifying, tracking, prioritizing, and remediating those vulnerabilities is often haphazard. It's a job every security program must do, but none of them enjoy it, and few do it well. The result is a patchwork of fixes applied to a tide of vulnerabilities, any one of which could be the single point of failure in an otherwise formidable defense.

But there's a better way, one rooted in data science, machine learning, and risk management. The result is a model that prioritizes—and even predicts—the vulnerabilities most likely to be exploited, most threatening to your organization, and most in need of remediation.

But before we dive in, it's important to lay a foundation of key terms we will use throughout this book. With so many approaches, tools, and systems available for vulnerability management, some common security terms are poorly defined, misused, or overly broad or specific. There is no playbook for vulnerability management—this is an emerging science. As such, we need to reestablish some of the terms you've likely heard before, as well as introduce new concepts that we'll return to again and again throughout the book.

Let's get started.

1.1 THE SECURITY CANON: FUNDAMENTAL CYBERSECURITY TERMINOLOGY

One of the more important distinctions we have to make is the difference between a vulnerability and a threat, a distinction that could have helped Equifax. Understanding the difference could save you countless hours and resources as you work to secure the software and hardware assets on your network.

There are billions of *vulnerabilities*, the flaws or weaknesses in assets that could result in a security breach or event. Vulnerabilities arise through a variety of vectors, from assets' design or implementation to the procedures or controls designed to secure them. While all bugs cause unintended effects in an asset, vulnerabilities are bugs that could be used maliciously.

However, a vulnerability is distinct from a *threat*, which is the potential for a specific actor to exploit, or take advantage of, a vulnerability. Tens of thousands of vulnerabilities are discovered each year, but only a small fraction become threats. A vulnerability becomes a threat when an exploit (code that compromises a vulnerability) is written. *Exploits* are usually programs, sometimes just commands, but always they are written in the same way all software is: with a little motivation and some creativity. As we saw in the case of Equifax, published exploit code that enables attackers to easily weaponize a vulnerability is a good indicator of the probability of an exploitation to take place.

An *exploitation* is the actual event of using an exploit to take advantage of a vulnerability. This is the threat becoming materialized, coming to pass. Not all exploitations will cause harm. If you exploit the server that hosts the PDF of a sandwich shop's lunch menu, you've gained access to data, sure. But that data was never confidential to begin with, and access to that machine can't cause much trouble (hopefully). Other exploitations are far more impactful, such as the Equifax Apache Struts exploitation above, causing millions of dollars in damage.

Different exploits and threats carry different levels of *risk*. According to ISO Guide 73, risk is the "combination of the probability of an event and its consequence." Some risk is unsystematic, caused by forces that are difficult to predict. Other risk is systematic and can be accurately predicted over time in aggregate. Risk management, as

1.1 THE SECURITY CANON: FUNDAMENTAL CYBERSECURITY TERMINOLOGY

defined by ISO Guide 73, is "the process of determining an acceptable level of risk, assessing the current level of risk, taking steps to reduce risk to the acceptable level, and maintaining that level of risk."

This is the basis of what we're going to explore in this book: How to assess, manage, and reduce your risk by mitigating the security vulnerabilities that pose the greatest threat to your assets.

Risk is the core of what security teams deal with, so this particular definition demands our attention. We prefer a slightly more philosophical formulation of risk, from Elroy Dimson of the London School of Economics: "Risk means more things can happen than will." At the core of risk is uncertainty, and uncertainty about attackers, events, systems, and ultimately the security of our systems is exactly what this book will help you understand and reduce.

If there was exactly one outcome, a 100% probability of you getting hit by a car going 60 miles per hour when you cross the road, then there wouldn't be any risk at all. You simply would not cross the road. But if the probability of the event drops below 100%, risk enters the equation. We process risk on a daily basis. Almost everything we do has some degree of uncertainty, and we come up with a tolerance to risk, or an amount of risk that we deem acceptable. There is a real, nonzero probability of being hit by a car when we cross roads, and the risk can vary based on the impact: Is the car pulling out of a parking spot at 1 mile per hour, or driving down a freeway at 100 miles per hour? Your risk when crossing the street is usually small, well mitigated by crosswalks, stoplights, and societal norms. The same is true in vulnerability management, and because the number of decisions vulnerability management requires is growing exponentially, we can no longer afford to make risk tolerance decisions unscientifically.

A better approach starts with data, and we need to understand the sources of data on existing vulnerabilities. The vulnerability assessments that scan and analyze assets for existing vulnerabilities and the vulnerability management efforts at mitigating relevant vulnerabilities both source data from several organizations and databases are detailed below.

1.1.1 Common Vulnerabilities and Exposures

Common vulnerabilities and exposures [3] is a list of known cybersecurity vulnerabilities, including unique ID numbers (commonly called CVEs) and standardized descriptions that facilitate information sharing among security professionals. It's maintained by MITRE and is free to use.

We will focus our research on discovered and disclosed vulnerabilities contained in the CVE list because CVEs are publicly tracked, readily available, extensive (although not exhaustive), and have become the de facto standard adopted by many other projects and products.

However, CVEs are neither comprehensive nor perfect. Many vulnerabilities are unknown, undisclosed, or otherwise have not been assigned a CVE ID. Furthermore, CVE listings are curated by humans, which makes them vulnerable to biases, errors, and omissions. Despite these challenges, the CVE list is a valuable community resource that greatly assists the otherwise untenable task of vulnerability management.

Figure 1.1 shows the volume of published CVEs by month. From the list's inception through January 1, 2018, over 120,000 CVE entries have been created. Of those, 21,136 are still in "reserved" status, meaning they have been allocated or reserved for use by a CVE Numbering Authority (CNA) or researcher, but the details have not yet been populated. Another 4,352 have been rejected for various reasons and eight are split out or labeled as unverifiable.

For all intents and purposes, each of these published CVEs represents a decision and potential action for vulnerability management programs. The criteria for those decisions may be simple in the

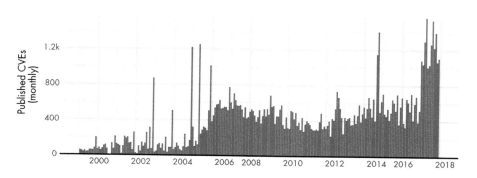

Figure 1.1 Monthly volume of published CVEs from 1999 through 2017. (© 2018 Kenna Security/Cyentia Institute [4]. Reprinted with permission.)

singular case (e.g., "Does that exist in our environment?") but prove to be quite difficult in the aggregate (e.g., "Where do we start?").

1.1.2 National Vulnerability Database

The National Vulnerability Database (NVD) is "the U.S. government repository of standards-based vulnerability management data" [5] that syncs with and enriches base CVE information, providing additional analysis, a database, and a search engine. NVD adds details by leveraging other community projects, including the CVSS, Common Weakness Enumeration (CWE), and Common Platform Enumeration (CPE).

1.1.3 Common Vulnerability Scoring System

CVSS [6] provides a process to capture the principal characteristics of a vulnerability and produce a numerical score that reflects its severity. CVSS was developed and is maintained by the Forum of Incident Response and Security Teams (FIRST).

1.1.4 Common Weakness Enumeration

CWE [7] provides a common language for describing software security weaknesses in architecture, design, or code. It was developed and is maintained by MITRE. Each piece of enrichment data offers potentially useful context for decisions. Basic remediation strategies may rely on CVSS alone, while others will factor in the type of vulnerability (CWE) along with the vendor and product and the exposure of the vulnerabilities across environments.

1.1.5 Common Platform Enumeration

CPE [8] provides a standard machine-readable format for encoding names of IT products, platforms, and vendors. It was developed at MITRE, but ongoing development and maintenance is now handled by the National Institute of Standards and Technology (NIST).

Each piece of enrichment data offers potentially useful context for making decisions. Basic remediation strategies may rely on CVSS alone, while others will factor in the type of vulnerability (CWE) along with the vendor and product and the exposure of the vulnerabilities across environments.

All of this data about vulnerabilities, however, might be useless if you don't have a strong understanding of which assets exist in your environment. As we'll discuss later in this book, asset discovery and management is a never-ending task. Your assets are constantly changing and evolving, and any understanding of what's at work in your environment is just a snapshot in time—your list of assets certainly changes month to month, and likely day to day or even hour to hour.

What's important is that you strive for the highest degree of completeness, meaning that you have the most current and accurate knowledge of as many assets as possible in your environment. That includes not only the obvious devices and software, but also the shadow IT and other assets that often operate under the radar. They too contain vulnerabilities that could pose a threat to your organization. Completeness is a measure of how many of the assets in your environment are included in your vulnerability management system so you have as much visibility as possible into your risk.

But as we saw with Equifax, simply being aware of vulnerabilities and assets is not enough. We need to know which vulnerabilities to prioritize and also how to measure our success in doing so.

1.2 SECURITY METRICS: THE NEW GUARD

In a perfect world, all vulnerabilities would be remediated as they were discovered, but unfortunately that doesn't happen in reality. The reality involves trade-offs between making the system more secure and rendering it unusable while it updates—sometimes causing disruptions in payment systems. It involves multiple groups of people with competing priorities: the security team that identifies the vulnerabilities, the IT operations or business owners that determine when the systems should be patched or a mitigating control is deployed, and the users of the systems themselves, internal or external. Users for whom security is at best a second-order benefit ("This new social network is so fun! And I hear it's secure."), and at worst, a cost center that looks bad on a balance sheet. With thousands of new vulnerabilities discovered every year multiplied across disparate assets, reality necessitates prioritization. It comes down to choosing a subset of vulnerabilities to focus on first.

It's tempting to go for overall accuracy—proportion decided correctly—but this can be misleading when so many vulnerabilities are never exploited. For example, if a company chooses to never remediate anything, that decision has an accuracy somewhere around 77%, seen in Figure 1.2. It might look like a good strategy on paper, but not so much in practice. The other mode of failure for the accuracy metric happens when maliciousness is rare, which is often the case in security. Let's say there's one machine with one vulnerability that's been exploited. In an environment with 100 machines with 10 vulnerabilities each, if every morning the vulnerability management director reported to the *chief information security officer* (CISO) that "we have no risky vulnerabilities across our entire network," then they would have an accuracy of 99.9%. Great score, but a useless metric here.

Instead of decision model accuracy, we will focus on the two concepts of coverage and efficiency.

Coverage measures the completeness of remediation. Of all vulnerabilities that should be remediated, what percentage was correctly identified for remediation? For example, if 100 vulnerabilities have existing exploits, and yet only 15 of those are remediated, the coverage of this prioritization strategy is 15%. The other 85% represents unremediated risk. Technically, coverage is the true positives divided by the sum of the true positives and false negatives.

Efficiency measures the precision of remediation. Of all vulnerabilities identified for remediation, what percentage should have been remediated? For example, if we remediate 100 vulnerabilities, yet only 15 of those are ever exploited, the efficiency of this prioritization strategy is 15%. The other 85% represents resources that may have been more productive elsewhere. Technically, efficiency is the true positives divided by the sum of the true positives and false positives.

Ideally, we'd love a remediation strategy that achieves 100% coverage and 100% efficiency. But in reality, a direct trade-off exists between the two. A strategy that prioritizes only the "really bad" CVEs

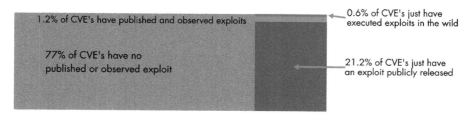

Figure 1.2 Comparison of CVEs with exploit code and/or observed exploits in the wild relative to all published CVEs. (© 2018 Kenna Security/Cyentia Institute [9]. Reprinted with permission.)

for remediation (i.e., CVSS 10) may have a good efficiency rating, but this comes at the cost of much lower coverage (many exploited vulnerabilities have a CVSS score of less than 10). Conversely, we could improve coverage by remediating CVSS 6 and above, but efficiency would drop due to chasing down CVEs that were never exploited.

Once you're set up to efficiently analyze the vulnerability data, a new world of cybersecurity metrics and concepts opens up.

Organizations will all end up in different places with respect to coverage and efficiency, as we discussed above, and they can take dramatically different routes to get there. It's useful, therefore, to understand remediation velocity, the measure of how long it takes to remediate a vulnerability.

Assume an organization observes 100 open vulnerabilities today (day zero) and manages to fix 10 of them on the same day, leaving 90 to live another day. The survival rate on day zero would be 90% with a 10% remediation rate.

As time passes and the organization continues to fix vulnerabilities, that proportion will continue to change. Tracking this change across all of the vulnerabilities will produce a curve like the one shown in Figure 1.3. From this curve you can observe that the overall half life of a vulnerability is 159 days. Beyond that, there's clearly a long-tail challenge that results in many vulnerabilities remaining open beyond one year.

Remediation time frames vary substantially. This is why we refer to remediation velocity; there's both a directional and a speed aspect to these lines. To quantitatively benchmark time frames, we can use several metrics from survival analysis, each of which gives a slightly different measure of remediation timelines:

- *Mean time to remediation* (MTTR) is the average amount of time it takes to close vulnerabilities;
- *Mean time to detection* (MTTD) is the average amount of time it takes to detect vulnerabilities;
- *Vulnerability half life* is the time required to close exactly 50% of open vulnerabilities;
- *Area under the survival curve* (AUC) represents live (open) vulnerabilities. A lower AUC means higher velocity. Figure 1.3 shows an example of a survival analysis curve.

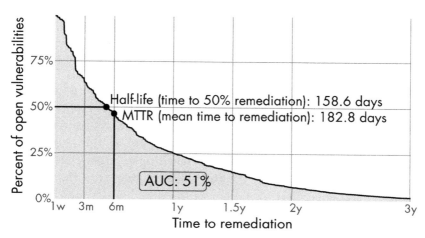

Figure 1.3 Graph depicting the survival analysis curve for vulnerability remediation timelines. (© 2019 Kenna Security/Cyentia Institute [10]. Reprinted with permission.)

While some organizations have a high velocity within a limited scope, others are playing the long game. Remediation capacity measures how many vulnerabilities you can remediate and how many high-risk vulnerabilities build up in your environment over time, using two primary metrics. Mean monthly close rate (MMCR) measures the proportion of all open vulnerabilities a firm can close within a given time frame. Vulnerability debt measures the net surplus or deficit of open high-risk vulnerabilities in the environment over time.

Think of MMCR as raw remediation capacity. To derive it, we calculate a ratio for the average number of open and closed vulnerabilities per month for each organization in our sample. On average, organizations remediate about one out of every 10 vulnerabilities in their environment within a given month.

Ultimately, these measures gauge the performance of your remediation within a program of *risk-based vulnerability management* (RBVM), a cybersecurity strategy in which you prioritize remediation of vulnerabilities according to the risks they pose to your organization.

RBVM means using threat intelligence to identify vulnerabilities attackers are planning to take action on externally. It means using intelligence to generate risk scores based on the likelihood of exploitation internally. It means using business context to determine the segments of your network where intrusion may be more damaging. It means combining all these elements into a measure of asset criticality. RBVM programs focus patching efforts on the vulnerabilities that are most likely to be exploited and reside on the most critical systems.

Note that the operative term here is risk-based, but in order to base our management decisions on risk, we need a set of metrics for measuring risk. We need those metrics to be universal, consistent, measurable, repeatable, and a slew of other semitechnical descriptions of a metric that are not to be taken lightly. In the end we'll use these metrics to construct machine learning models that automate some of the heavy lifting for us. But these metrics, or objective functions, have to be as clean and clear as the data being processed.

At the turn of the twenty-first century, at the height of the hype cycle of big data, machine learning practitioners and political pundits alike have come to agree that when it comes to data, garbage in equals garbage out. One piece that's missing from this kind of thinking is that data is only valuable in a particular context. We don't care much about the precision of the temperature in Antarctica on any given day. For our purposes, "extremely cold" or "warmer than Chicago" is good enough. To a climatologist, or someone studying phase-transitions of metals, a decimal rounding error might be catastrophic. Machines are context-aware only in the most notional of senses—it depends on what we call context. Pay attention to the metrics we use to build these models. A well-crafted metric will define what good data looks like, and a good set of objective functions mathematically describes the context that all models need to be useful.

RBVM paves a pathway to modern vulnerability management, an orderly, systematic, and data-driven approach to enterprise vulnerability management. It leverages full visibility into a technology stack to target the riskiest vulnerabilities, enabling companies to adhere to designated service-level agreements (SLAs), respond to threats rapidly and have meaningful discussions about organizational risk tolerance.

In a traditional vulnerability management system, like the method Equifax was using in 2017, nobody is using clear, undisputed data that gives certainty into which actions matter. Modern vulnerability management programs make order from this chaos.

The hallmarks of a modern vulnerability management program are a consistent, systematic, and ongoing method to discover risk within the enterprise environment. It's a data-driven approach that helps companies align their security goals with the actions they can take.

Developing a modern vulnerability management program isn't like flipping a switch. It's an evolution, with several steps along the way. Each chapter in the remainder of this book will guide you step by step on how to design and implement such a program.

References

[1] U.S. House of Representatives Committee on Oversight and Government Reform, The Equifax Data Breach, *Majority Staff Report*, 115th Congress, December 2018, https://republicans-oversight.house.gov/wp-content/uploads/2018/12/Equifax-Report.pdf.

[2] Biasini, N., "Content-Type: Malicious—New Apache Struts2 0-day Under Attack," *Cisco Talos Blog*, March 8, 2017, https://blog.talosintelligence.com/2017/03/apache-0-day-exploited.html.

[3] Common Vulnerabilities and Exposures (CVE) Program, https://cve.mitre.org/index.html.

[4] Kenna Security and The Cyentia Institute, *Prioritization to Prediction: Analyzing Vulnerability Remediation Strategies*, 2018, https://learn-cloudsecurity.cisco.com/kenna-resources/kenna/prioritization-to-prediction-volume-1.

[5] National Vulnerability Database, https://nvd.nist.gov/.

[6] Common Vulnerability Scoring System (CVSS), https://www.first.org/cvss/.

[7] Common Weakness Enumeration (CWE), https://cwe.mitre.org/.

[8] Common Platform Enumeration (CPE), https://cpe.mitre.org/.

[9] Kenna Security and The Cyentia Institute, *Prioritization to Prediction: Analyzing Vulnerability Remediation Strategies*, 2018, https://learn-cloudsecurity.cisco.com/kenna-resources/kenna/prioritization-to-prediction-volume-1.

[10] Kenna Security and The Cyentia Institute, *Prioritization to Prediction Volume 4: Measuring What Matters in Remediation*, 2019, https://learn-cloudsecurity.cisco.com/kenna-resources/kenna/prioritization-to-prediction-volume-4.

2

DATA SCIENCE TO DEFINE RISK

The concept of risk management hasn't changed much over the millennia. Some of the earliest examples include the Babylonians, who assessed risk in sea voyages. Similar risk management practices continued for thousands of years until they were formalized as operations research during World War II, helping the U.S. military assess vulnerabilities in airplanes. After the war, managing risk and projecting rewards became business imperatives and lessons every MBA learns.

Cybersecurity, however, is different. While the concept of risk management hasn't changed, the approaches to it have. In cybersecurity, the scale of vulnerabilities is orders of magnitude greater than in other disciplines. It requires a different approach, one rooted in the history of risk management but one that also leverages machine learning and data science to define, understand, and mitigate risk.

The problem requires that we build a decision engine for vulnerability management at scale, and the bulk of this chapter will cover what you need to do so. But first, let's look at the history of risk management to understand what it is we're trying to achieve and why we examine risk to make decisions.

2.1 RISK MANAGEMENT HISTORY AND CHALLENGES

One of the earliest forms of risk management was a proto-insurance called bottomry, which was designed to limit the risks of seafaring.

Dating back to at least ancient Babylonia and described in the Code of Hammurabi, the practice was borrowed by the Greeks and then the Romans. Shipping goods by sea had inherent risks—weather caused delays, ships sank, crews mutinied, pirates plundered, and prices changed before ships arrived in port.

Bottomry was a high-interest loan that merchants took out to fund their journeys at sea and cover repairs or handle other emergencies the ship might encounter, using the "bottom" of the ship as collateral. If the ship arrived at its destination successfully, the master of the ship would repay the loan with interest. If the voyage ended in disaster, the loan was void and the borrower owed the lender nothing. If the voyage was successful but the master of the ship didn't pay back the loan with interest on time, the lender could seize the ship.

Often called a precursor to maritime insurance, bottomry has also been described as a "futures contract" in which the insurer "has bought an option on the venture's final value." Interest rates were high, and the creditors issuing the loans saw the high interest rates as compensation for risk [1]. Interest rates would rise during stormy seasons or for unreliable borrowers [2].

Approaches to managing risk evolved but remained largely the same for the next 2,000 years. The first major shift came during World War II with the first formalization of operations research. Mitigating risk and addressing vulnerabilities became more complex and the scale of the risk required a more systematic approach to management.

2.1.1 The Birth of Operations Research

Operations research is most often traced back to statistician Abraham Wald and his work with the Statistical Research Group (SRG), which comprised the brightest mathematical minds available to aid the American effort in World War II.

At the time, the U.S. military was seeking ways to reduce the loss of aircraft in combat. Ultimately, the planes needed better armor. The only question was how much armor and where to place it. One option would be to simply reinforce armor across the entire aircraft. But too much armor weighed down the plane, burning fuel more quickly and limiting its maneuverability. Another option was to apply armor only to the most vulnerable parts of the plane. But which parts were most vulnerable?

2.1 RISK MANAGEMENT HISTORY AND CHALLENGES

The military gathered data. When planes returned to the U.S. after battles in Europe, the military examined them and noted which parts of the plane had the most bullet holes, measuring bullet holes per square foot on the engine, the fuel tank, the fuselage, and other parts of the plane. The distribution of bullet holes looked similar to that of the plane in Figure 2.1.

It turned out that the fuselage and the other parts of the plane had nearly twice as many bullet holes as the engine. Since those areas had taken the most fire and were where the crew and key equipment were located, the military decided to reinforce the armor on the fuselage and other parts. They turned to Abraham Wald at the SRG to figure out how much armor would be necessary.

What they received in reply was a reframing of their assumptions about vulnerability. The armor, Wald argued, belonged not in the area with the most bullet holes, but the area with the fewest: the engines. The planes that had returned had managed to do so despite all those bullet holes in the fuselage because they had taken fewer shots to the engines.

Upon seeing the data, Wald understood that it was not statistically accurate and was based on one major assumption: that the

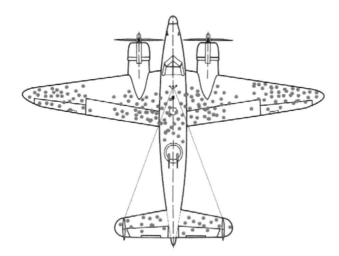

Figure 2.1 Representation typical of bullet hole distribution on a plane returned from battle during World War II. (Reprinted under a Creative Commons Attribution-Share Alike 4.0 International Public License, © 2005, 2016, 2021 [3].)

returning aircraft were a representative sample of all aircraft. Unfortunately, the data left out a crucial component: the planes that hadn't made it back from battle. Data on the downed aircraft was not available, but Wald deduced that the engines of those planes were likely littered with bullet holes.

Wald set to work and delivered a memorandum to the military that estimated the probability that any additional hit would take down an aircraft after it had already been hit, the vulnerability of different sections of an airplane, and a way to estimate the damage on the planes that never returned [4].

In this case, the scale of the problem had become big enough that looking at just one plane or just the planes that returned wasn't sufficient. What the military needed was to look at the statistical population of planes to discover the most critical vulnerabilities. It needed a systematic way to observe and mitigate the greatest risks. Wald's solution and recommendations not only saved countless lives during World War II, but was used by the Navy and Air Force for decades afterward.

2.1.2 The Scale of Cybersecurity

Following World War II, similar principles were applied to business. In addition to prioritizing vulnerabilities, prediction too became systematized. Engineer and statistician Genichi Teguchi developed what are now known as the Teguchi methods to improve manufacturing quality. Part of his principles is understanding the variables and parameters in design that have the most impact on performance in the final product. MBA programs now teach businesspeople the language of risk-to-reward ratios, the probability of success, the expected return on investment, the time value of money, and so on, each of which has been systematized to spot opportunities and mitigate risk.

Which brings us to security. While operations research and financial risk and reward have long been codified, security presents different challenges. One of the more immediate ones is that security lacks the decision support that money provides.

Ancient world merchants raised interest rates with the level of risk of shipping goods by sea. MBAs chart recommended actions based on the expected returns. But security is a loss leader. The calculation is more about the probability of loss and estimating the loss function.

You're trying to prove that your company saves money, time, and resources by an event not happening, which is difficult to do. You still have to make decisions, but instead of basing them on financial gain or loss, they must be based on metrics you're creating yourself.

The other major challenge facing cybersecurity is the vast—and rapidly growing—scale of vulnerabilities. While it's relatively simple to understand the risks facing a single ship leaving port in the ancient world or even the weaknesses in World War II aircraft, cybersecurity is much more complex. The number of vulnerabilities continues to surge month after month, assets continue to multiply in organizations, and exploits continue to grow more damaging and pervasive. The vulnerability management strategies of the past need to scale to the breadth of the challenge today.

In 2014, cybersecurity and risk management expert Dan Geer gave a keynote at Black Hat USA titled "Cybersecurity as Realpolitik" [5]. He led off by acknowledging the scale of the problem:

> I wish that I could tell you that it is still possible for one person to hold the big picture firmly in their mind's eye, to track everything important that is going on in our field, to make few if any sins of omission. It is not possible; that phase passed sometime in the last six years. I have certainly tried to keep up but I would be less than candid if I were not to say that I know that I am not keeping up, not even keeping up with what is going on in my own country much less all countries. Not only has cybersecurity reached the highest levels of attention, it has spread into nearly every corner. If area is the product of height and width, then the footprint of cybersecurity has surpassed the grasp of any one of us.

In fact, over the six-year period Geer mentions, security professionals had been seeking new ways to manage a set of risks that was rapidly growing beyond their ability to comprehend or to manage. Geer addresses this issue later in his keynote when he notes the vast numbers of vulnerabilities:

> In a May article in *The Atlantic* [6], Bruce Schneier asked a cogent first-principles question: Are vulnerabilities in software dense or sparse? If they are sparse, then every one you find and fix meaningfully lowers the number of avenues of attack that are extant. If they are dense, then finding and fixing one more is essentially irrelevant to security and a waste of the resources

spent finding it. Six-take-away-one is a 15% improvement. Six-thousand-take-away-one has no detectable value.

This is the conundrum the security community faces. But it's important to keep in mind the distinction between a vulnerability and an exploit. Fortunately, the vast majority of reported vulnerabilities aren't used by hackers—77% never have exploits developed and less than 2% are actively used in attacks. Fixing one vulnerability out of 6,000 does have value if you've fixed the right one—one that poses the most risk to your organization. If less than 2% of vulnerabilities are actively used in attacks, among Geer's example of 6,000, you might only have to remediate 120 to make a measurable difference. The trick is, you have to know which vulnerabilities to mitigate, and from among a pool that could far exceed 6,000 and grows by the day.

2.1.3 Origins of the Risk-Based Approach to Vulnerability Management

RBVM can be traced back to 1999 with the advent of the CVSS. CVEs were one of the first ways to measure and identify vulnerabilities and bring all security practitioners onto the same page. At the time there was nothing better—it was revelatory and even offered a good amount of efficacy.

Around the same time, compliance and audit rules sprang up, introducing other metrics and taking a business approach to regulating cybersecurity. The goal was to manage risk, but the guardrails were often put in place using incomplete, inaccurate, or just plain inefficient metrics. This leaves companies bowling with bumpers that are far too wide.

The Payment Card Industry Data Security Standard (PCI DSS), for example, used to instruct companies to fix every vulnerability with a CVSS score of 4 or above (on a scale from 0 to 10).[1] Of the vulnerabilities listed in CVSS, 86% score a 4 or higher, meaning the guideline was essentially to fix nearly everything. In a sense, these compliance policies were a very raw, finger-to-the-sky version of risk-based vulnerability management. They set a risk threshold, determined time

---

1. PCI DSS has since added some flexibility with version 3.2.1, which requires companies to fix all vulnerabilities with a CVSS score of 4 or above for external scans, and all "high risk" vulnerabilities, as defined in the PCI DSS Requirement 6.1, for internal scans. Users can also use a scoring system outside of CVSS.

2.1 RISK MANAGEMENT HISTORY AND CHALLENGES 21

frames, and made them actionable. As we'll see throughout this book, the intent was well meaning but the data and the data-driven methodology was missing. To claim "this payment system is safe if it has no vulnerabilities with a CVSS score above 4" is not the same as "all global 2000 enterprises must remediate 86% of their vulnerabilities within 90 days." The policy prescription had unintended consequences: When everything is a priority, nothing is.

This is the demoralizing task cybersecurity practitioners faced in the early 2000s. No business can remediate 86% of its vulnerabilities. No business can afford to. Even if it can, doing so takes time, and it'll be exposed to some risks for years while others are quickly remediated—is that safe? Which ones should it remediate first? Organizations must make much tighter decisions based on the specific context of their business and the vulnerabilities that pose the greatest risk to ongoing operations.

Around the time PCI was enacted in 2005, the precursors to RBVM started to take shape. At that time, security practitioners needed more efficient ways to identify vulnerabilities. Up until then, manual assessments were the primary method, one that cost a lot of time and effort with minimal reward. Teams might find some of the many vulnerabilities, but ultimately didn't have visibility into risk across the entire organization. Automated tools and scanners sprang up to fill this need.

Unfortunately, this new data did little to assuage concerns about risk. The scanners generated reports filled with bad news: Organizations had vulnerabilities everywhere. Scanners ensured organizations were more aware of the scope of the problem they faced, but offered no strategy for facing it. Security practitioners created systems around these tools to measure the effectiveness of their remediation. They would push out patches for vulnerabilities the scanners had uncovered and then run a new report to determine their progress.

The result was often depressing and demoralizing: No matter how quickly companies patched vulnerabilities, they barely put a dent in the ever-growing number—more and more vulnerabilities were discovered every day and security teams couldn't keep up with the sheer number of them.

Also around the mid-2000s, some network mapping tools started adding vulnerability data to their products so organizations could start to understand exactly what parts of their network vulnerabilities might impact. By examining firewall and router rules, access control lists (ACLs), and other factors at the network layer, these tools offered organizations a better sense of which vulnerabilities mattered most to them.

The pieces for a cybersecurity risk-management decision engine were starting to come together. In 2010, NIST released the Security Content Automation Protocol (SCAP) both to commonly describe vulnerabilities across multiple tools and then to aggregate and prioritize vulnerabilities, even introducing a standard for confirmation scoring. It was a way to prioritize configuration issues as well as application security weaknesses. SCAP ensured that everyone was using the same metrics to describe the same vulnerabilities, and thus served as the seed for automating vulnerability management.

Not long after, Anton Chuvakin of Gartner [7] confirmed that "vulnerability prioritization for remediation presents the critical problem to many organizations operating the scanning tools. The volume of data from enterprise scanning for vulnerabilities and configuration weaknesses is growing due to both additional network segments and new asset types."

At the end of 2015, Chuvakin and Augusto Barros released *A Guidance Framework for Developing and Implementing Vulnerability Management* [8], a piece that set out several of the concepts critical to vulnerability management, including maximum patching speed and setting context based on SLAs, as well as a process workflow that set the stage for automation.

By 2016, Gartner was examining different models of vulnerability remediation:

- *Asset-centric:* Focused on the business criticality, value of, and exposure of an asset, with an approach of gradual risk reduction;
- *Vulnerability-centric:* Focused on the criticality of a vulnerability, also with an approach of gradual risk reduction;
- *Threat-centric:* Focused on vulnerabilities actively targeted by malware, ransomware, exploit kits, and threat actors in the wild, with an approach geared toward imminent threat elimination.

All of these point to the search for an answer to a simple question: What's the likelihood this vulnerability will be exploited, and if it is, what will be the impact to a particular organization?

Each model has multiple layers as well. In the threat-centric view, for example, you must consider that there are targets of opportunity—vulnerabilities that are exposed and likely to be exploited—but there are also targeted attacks—a bad actor coming after you because of who you are. Targeted attacks are a different and often much more expensive situation to deal with. The threat-centric view is ultimately the one we want to focus on, but there are multiple layers to operationalizing the approach.

In January 2018, Gartner started to use the descriptor "risk-based," describing a risk-based approach to manage enterprise vulnerabilities in highly regulated industries. As security professionals, we're paid to understand risk, the cost to mitigate that risk, and whether that cost is justified. Business context should always be taken into account: What is your organization's risk appetite—are you a too-big-to-fail bank or a mom-and-pop down the street? Vertical size matters, what you're protecting matters, how your industry is regulated matters. All of this context about your business and what it is you do are critical factors to consider in effective vulnerability management.

In his 2014 Black Hat keynote [5], Dan Geer argues that if we define risk as "more things can happen than will," then what is relevant is either playing offense or getting out of the line of fire altogether. The latter of which involves stepping back from dependencies on digital goods and services—those "who quickly depend on every new thing are effectively risk seeking," Geer notes—something that may be possible for individuals, but is increasingly impossible for businesses. Offense, in this case, amounts to truly understanding the risks in the larger context of your business and prioritizing and predicting which of the vulnerabilities that can be exploited pose the greatest risk to your organization.

We can take Geer's claim about risk-seeking businesses even further. Without an a priori way to measure the risk that a new technology or behavior introduces into the business, without some forecast of the future, businesses aren't making rational decisions. Instead, they're choosing to ignore a set of risks, sometimes existential, sometimes marginal, always present.

That's our aim in this book. To create a decision-support engine to make better choices based on risk that keep your organization secure. In essence, it's understanding which bullet holes will down your plane, and thus where to reinforce your armor.

References

[1] Wang, D., "How Maritime Insurance Helped Build Ancient Rome," Flexport, January 29, 2016, https://www.flexport.com/blog/maritime-insurance-in-ancient-rome/.

[2] Priceonomics, "How Maritime Insurance Built Ancient Rome," Priceonomics.com, March 18, 2016, https://priceonomics.com/how-maritime-insurance-built-ancient-rome/.

[3] Illustration of hypothetical damage pattern on a WW2 bomber. Based on a not-illustrated report by Abraham Wald (1943), picture concept by Cameron Moll (2005, claimed on Twitter and credited by Mother Jones), new version by McGeddon based on a Lockheed PV-1 Ventura drawing (2016), vector file by Martin Grandjean (2021).

[4] Caselman, B., "The Legend of Abraham Wald," *American Mathematical Society*, June 2016, http://www.ams.org/publicoutreach/feature-column/fc-2016-06.

[5] Geer, D., "Cybersecurity as Realpolitik," Black Hat USA 2014, http://geer.tinho.net/geer.blackhat.6viii14.txt.

[6] Schneier, B., "Should U.S. Hackers Fix Cybersecurity Holes or Exploit Them?" The Atlantic, May 19, 2014, https://www.theatlantic.com/technology/archive/2014/05/should-hackers-fix-cybersecurity-holes-or-exploit-them/371197/.

[7] Chuvakin, A., "On Vulnerability Prioritization and Scoring," Gartner, October 6, 2011, https://blogs.gartner.com/anton-chuvakin/2011/10/06/on-vulnerability-prioritization-and-scoring/.

[8] Barros, A., and Chuvakin, A., *A Guidance Framework for Developing and Implementing Vulnerability Management, Gartner*, November 17, 2015, https://www.gartner.com/en/documents/3169217.

3

DECISION SUPPORT: TAPPING MATHEMATICAL MODELS AND MACHINE LEARNING

Exploits are a unique adversary, especially compared to the risk of pirates and enemy fire we've discussed.

The threats facing Babylonian ships—pirates, weather, mutiny, and so on—can be mapped along a bell curve. At the edges are the extremes: the especially destructive storms, the strongest and most ruthless pirates on one side, and the toothless attackers and mild drizzles on the other. But those instances are exceedingly rare. For the most part, threats will be about average, the healthy middle of the curve.

The success of enemy soldiers is also governed by physical constraints like strength and hand-eye coordination, as well as the accuracy and destructiveness of their weapons. There are few deadeye sharpshooters and few with lousy aim. Most are about average.

Cybersecurity exploits, however, operate differently. Instead of a bell curve, they fall along a power-law distribution. Of the thousands of vulnerabilities, the vast majority pose no threat. A small number of vulnerabilities are ever exploited. And an even smaller portion of those exploits cause outsized destruction.

This distinction completely changes the method of defense for different attacks. Ships and planes can get away with armor and defenses that are fairly uniform because most attacks will come from the heart of the bell curve. Certainly enemies or weapons at either end of the curve could pierce those defenses, but encountering them is rare. So for most situations, the typical armor will hold up.

Not so in the cybersecurity realm. An especially damaging exploit isn't limited by physical location and could affect every user of the vulnerable asset. To make matters more complicated, networks, applications, users, and other factors are continually evolving and changing, meaning a defense that worked at one point in time might no longer work the same way at a later point.

Designing planes with the same defenses saved lives during World War II. But as David L. Bibighaus points out [1], introducing assets into an IT environment that are all secured and hardened the same way is a vulnerability, since unlike bullets, the exploits can evolve rapidly, and much like with evolution, a system with homogeneous strengths also has homogeneous weaknesses. If the exploit evolves in the right way, it can take down all the devices instead of just a few.

No one can tell at the outset which vulnerabilities will be among the few to be exploited or anticipate the ways those exploits will evolve and affect their organization's systems, especially as those systems also evolve.

In this kind of threat environment, we need a way to constantly evaluate the probability of various risks and pinpoint the vulnerabilities most in need of remediation. We need to continually refocus our attention on the most probable attack vectors as the threats and systems change in a complex environment where there are many unknowns. What we need is a decision support system.

Behind every good decision support system are mathematical models and machine learning algorithms that help us efficiently make stronger decisions despite uncertainty, randomness, and change.

3.1 MATHEMATICAL MODELING

In this section, we're going to examine five methods of mathematically representing the problem we face and gain insight into how we can best approach it. These models will help account for the scale of the number of variables and interdependencies we'll encounter, help us quickly take action using limited information, and account for randomness, uncertainty, and sentient, rational attackers.

3.1.1 Mathematical Scale

One of the oldest river-crossing puzzles, dating to at least the ninth century, is the wolf, goat, and cabbage problem. A farmer must transport a wolf, goat, and cabbage across the river, but he can only carry himself and one other on each trip. But if he leaves the wolf alone with the goat, or the goat alone with the cabbage on either side of the river, the former would eat the latter in either case. How can he move all three to the other side of the river?

It's a simple problem, and one you've likely solved before. It's also fairly simple to construct a mathematical proof for the fewest number of trips. But the problem grows more complex with the number of objects. As soon as we move from three to four or five in the same style of problem, the same proof no longer applies due to the interdependencies introduced. This is an example of mathematical scale. As soon as the number of elements exceeds two or three, the simple proof that was possible no longer applies.

There's an analogy there to algorithmic complexity. Algorithms don't scale linearly if you have more data. What worked for a smaller problem won't scale to a larger one.

Think first in terms of physical security. Making one risk management decision is tractable. For example, should I install a floodlight in front of my garage door? As the complexity and interdependencies increase, the problem becomes much more difficult. For example, how do I secure the neighborhood? Suddenly there are multiple interdependencies and changing one element leads to changes in another.

This kind of mathematical scale is illustrated in concepts like P versus NP and wicked problems.

P versus NP (polynomial versus nondeterministic polynomial) refers to a theoretical question presented in 1971 by Stephen Cook [2] concerning mathematical problems that are easy to solve (P type) as opposed to problems that are difficult to solve (NP type).

Any P type problem can be solved in polynomial time. A P type problem is a polynomial in the number of bits that it takes to describe the instance of the problem at hand. An example of a P type problem is finding the way from point A to point B on a map.

An NP type problem requires vastly more time to solve than it takes to describe the problem. An example of an NP type problem is breaking a 128-bit digital cipher. The P versus NP question is important in communications because it may ultimately determine the effectiveness (or ineffectiveness) of digital encryption methods.

An NP problem defies any brute-force approach, because finding the correct solution would take trillions of years or longer even if all the supercomputers in the world were put to the task. Some mathematicians believe that this obstacle can be surmounted by building a computer capable of trying every possible solution to a problem simultaneously. This hypothesis is called P equals NP. Others believe that such a computer cannot be developed (P is not equal to NP). If it turns out that P equals NP, then it will become possible to crack the key to any digital cipher regardless of its complexity, thus rendering all digital encryption methods worthless.

Complex interdependencies are also conceptualized in wicked problems, where solving one aspect of the problem creates other problems. A wicked problem can result not just by changing elements, but because of incomplete information. Every solution creates greater complexity and uncovers new problems. In these situations, there will inevitably be trade-offs.

One example that we'll cover in more detail in the next chapter is achieving coverage versus efficiency when remediating vulnerabilities. Coverage refers to your ability to fix everything that matters, to remediate as many exploited or high-risk vulnerabilities as you can. Efficiency is about the precision of your efforts; in this case, the percentage of the vulnerabilities you've remediated that were actually high risk. But achieving 100% coverage will decrease your efficiency since most of the vulnerabilities remediated won't be high risk. Pursuing efficiency at all costs will likewise reduce your coverage. It's always a balance and a trade-off.

By way of example, a financial services company knew its systems were rife with vulnerabilities but had not quantified the risk. A scan found 220 vulnerabilities per machine. Recognizing the scale of the problem, the company aggressively remediated those vulnerabilities, incentivizing employees by attaching bonuses to progress. But the result was the opposite of what the company expected: Its risk started rising, not falling.

By focusing on the problem of fixing vulnerabilities, the company discovered a new problem: Not all vulnerabilities are risky. It revamped its initiative to incentivize employees to address only the most serious vulnerabilities, aligning bonuses based on risk scores—the lower, the better. It didn't matter which vulnerabilities they remediated as long as they took care of the most important ones at the top of the list. But this created yet another issue. If the last day of the quarter happened to be the day after Patch Tuesday, when a bunch of new risks were revealed, risk scores would rise and could cause people to lose bonuses. So the employees would delay remediation to keep their risk score low.

Ultimately what ended up mattering more than risk was adherence to risk tolerance timelines and SLAs, incentivizing based on speed. But the complexity of the problem is clear: Each solution revealed a completely new problem.

As we build a decision support system, we have to keep in mind the mathematical scale of the problem and how unwieldy it can become. But there are ways to tame that complexity.

3.1.2 Statistics

The two main branches of statistics are descriptive and inferential. Descriptive statistics focuses on data that we know, while inferential statistics aims to extrapolate what we don't know from what we do [3]. Given a sample of data, descriptive statistics will analyze the features of that sample, while inferential statistics infers the features of the larger population of which the sample is a part.

The need for statistical inferences arises from the impossibility of collecting all data. Election pollsters, for instance, can't survey every voter for every poll. It would be too expensive and time consuming to be practical, but more significantly, in the time it would take to speak to every voter, some voters already surveyed might change their minds, new voters will have registered, and any number of other variables will have changed. It's not merely a matter of running down a list of names and recording responses. It's an ever-evolving complex system with various interdependencies. So instead, pollsters survey a representative sample of voters over a defined period of time and infer the preferences of the wider electorate based on the sample data collected.

Vulnerability assessment is not too different. Vendors need to write signatures for scans, by the time a scan of a sufficiently large network is complete, portions of the network may have changed, new vulnerabilities may have come out, what the business considers critical may have changed. It gets worse on the threat-measurement side. Methods don't exist for detecting every attack, we aren't aware of all of the vulnerabilities or exploits currently in use today, and nefarious groups go to great lengths to conceal some of these. In a sense it is actually impossible to collect all the data necessary to optimally solve vulnerability management.

As we've seen in recent presidential elections, polls can appear to be far off the final election results. One reason is that every poll is merely a snapshot in time and might be obsolete as soon as it's reported. There is always a margin of error, and the accuracy of the polling depends in large part on weighting the right variables. This kind of confidence interval expresses how closely the sample approximates the larger population. Larger samples sizes have smaller margins of error.

Successfully approximating populations depends in large part on collecting a strong sample and understanding how to weight the data. Doing so allows us to create models that help us deduce the larger cybersecurity threat landscape based on the data we have available.

In many ways, it functions similarly to epidemiology. According to the U.S. Centers for Disease Control and Prevention (CDC), epidemiology is "the study of the distribution and determinants of health-related states or events in specified populations, and the application of this study to the control of health problems" [4].

If you swap out "health-related states or events" for vulnerabilities and exploits, "specified populations" for a network and IT environment, and "health problems" for cybersecurity threats, it's in many ways a direct match.

Epidemiology allows us to find patterns and complex interdependencies within specific groups, identifying causalities that allow intervention and remediation. Applied to vulnerability management, we can use statistics to better understand why some organizations or industries approach security differently so we can measure and design interventions accordingly.

Organizations are often able to react to a particular vulnerability or event and decide what to do about it from a technical perspective. But understanding the statistical causes or statistical correlations across populations is much more difficult. Especially because, unlike the movement of a virus through a population, in cybersecurity both the attacker and defender are sentient. Both are thinking rationally, making predictions, and taking action.

It's important to remember that epidemiologists are not the doctors treating patients; they're simply approximating populations and making recommendations as a whole, hoping those recommendations are put into action. When creating statistical models, both in epidemiology and cybersecurity, you want a model that's only as complex as necessary. If the model grows too complex, it's more difficult to identify what is causal and how to intervene. Statistical models in cybersecurity must be actionable.

For example, if we develop a model that includes 400 different variables, and 300 of them come from closed sources that we're purchasing and scanning the dark web for, IT operations teams might be less likely to act on any predictions coming from that model because they couldn't explain why they were taking that action. On the flip side, if we had a model that was 95% precise, but only included 10 variables, IT operations might more intuitively understand the prediction and recommendation, leading to action. Always consider whether the model will be operationalized and acted upon.

Statistics can also assist in calculating the return on investment (ROI) of specific actions taken. Mass scanning of the internet and of organizations is a relatively recent tactic, but one that has accumulated large data sets around specific industries, like financial services and healthcare, and around the entire publicly accessible internet. Based on that data, we can test whether certain security protocols are worth the ROI.

Most importantly, it's not enough to simply analyze the data already available. To do so will be to fall behind. Which is why predictive analytics are so important. You also need continuous, always-on data even if you've built a statistical model that gets the causality right. Because the causality right now might not be the causality tomorrow. In epidemiology, pathogens evolve through natural selection, while in cybersecurity, they evolve in analogous methods through predator-prey dynamics. It requires that you continually gather data to reevaluate and adapt as the system shifts. Any model needs to be real-time and reactive.

3.1.3 Game Theory

The predator-prey dynamics that involve exploits and threats can be better understood and acted upon using game theory. Game theory is an applied field of mathematics developed as a means of computing optimal strategy sets for rational actors. In essence, game theory applies mathematical models to decision-making.

In cybersecurity, both attackers and defenders are intelligent, rational decision-makers, and adjust strategies and tactics at any time in response to incentives, actions, and anticipated actions. In this complex and chaotic environment, game theory can help us make sense of and predict future actions.

A common application is to analyze adversaries locked in a repeated game, with the outcome of one affecting constraints on the subsequent games. Repeated games assume that players will have to take into account the impact of their current strategy on the future actions of other players; this is sometimes called their reputation. In web security, the analogue is clear cut: Hackers are choosing which assets to exploit, and each one of us wants to decide which assets and vulnerabilities to fix first.

As a security professional, your job isn't to make one decision to stop one attack. It's devising a strategy to defend against attacks coming every hour indefinitely. In a RBVM program, a snapshot of the data shows vulnerabilities, assets, and where the most risk resides. But in reality, the risk is constantly changing and rebalancing, and defenders have to react to it. The guidance you give them on how to react, specifically the SLAs governing the amount of time they have to fix vulnerabilities, changes based on the fact that they have to do it consistently for a year.

In a simple game scenario, the 1% of vulnerabilities that are riskiest should be fixed as fast as possible. But if during that time period, the game resets, and you might have a new vulnerability arise, you might have a new attack. So, you have to balance the real-time nature of the vulnerabilities and exploits with the time it takes to respond with your limited capacity.

This is why concepts like capacity and efficiency (which we'll discuss in the next chapter) are so important to your efforts. Your strategy needs to take into account that you're converging on an equilibrium as part of an infinitely repeating game, rather than basing it on the data you've analyzed to begin with as you would in a simple game.

For example, one common security strategy is to map a network topology by analyzing every path and examining the network packets sent by each host. The goal is to understand spillover effects of a compromised asset. But if you zoom out and start to think about the repeated game, the interconnections between each machine matter much less.

In much of game theory, the best defense against an attacker, especially one about which there is little information, is dynamic. This is because attackers randomize their attack patterns and paths. As the number of attack paths grows, the value of a particular asset itself matters more than its effect on others. Simply put, over a long enough period of time, attackers will try every path. Trying to cut off one link when you know the attackers will take advantage of the others is a less effective strategy in a repeated game. You're not playing against one attacker one time. You're playing against multiple attackers and multiple types of attackers who continually evolve their strategies. Those changes and differences will also alter the importance of different assets, which in turn affects the importance of different links in a network.

All this is to say that a map of network topology doesn't need to be factored into a data model when choosing which assets to defend. This does not imply that we should ignore the topology of networks, but it does mean that we need to think critically about whether including all the data we have is the best approach to reducing risk. If the paths don't matter, then it's a question about which assets are the most important and which vulnerabilities are most likely to actually be exploited if every path is taken.

When you have an attacker and defender constantly interacting over an infinite period of time, you can begin to model the different incentives for each party. Each party has incomplete information about how the others will interact. Repeated game models allow security teams to anticipate future actions based on previous threats, thinking ahead not only to an attacker's next move, but how the security team's next move might influence future attacks. This kind of modeling allows the defender to predict the strategies used against them and take requisite precautions.

Unfortunately, the incentives are often aligned in favor of the attacker. The cost of a wrong move by the defender could result in a data breach, lost revenue, and other consequences. Attackers pay a far lower price for being wrong. Once they've developed an exploit, they can simply spray and pray. Even if they're wrong countless times, they only have to be successful once to cause damage.

3.1.4 Stochastic Processes

If you want to understand the probability of future outcomes, stochastic processes lend a helpful model. The general idea is that you can predict outcomes based on a number of observations at various points in time. The more observations you make, the more likely you are able to predict a later outcome. Each observation acts as a random variable that adds a certain probability of various outcomes. Using stochastic processes, we can anticipate the risk of exploits and attacks based on incomplete information and observations. Stochastic processes are commonly used in situations, like vulnerability management, where practitioners must account for randomness of events.

There are two main approaches to stochastic processes: analytical models and empirical models. Analytical models are based on mathematical reasoning, while empirical models leverage observed experience. Both can be useful, and there are various models that take both into account.

The random walk is the most basic stochastic process, and is depicted in Figure 3.1. You start at zero, and then with probability 50%, you go to one, probability 50%, negative one. And then you go up one down one in a random sequence.

This is the baseline for financial markets. It's not a straight line, and there's a completely random chance of going up and down. It mathematically models the time series component and probabilistic component, even if that component washes out to zero over a long period of time. But in finance, no one is actually focused on a long time frame. It's about buying before the price rises and selling before it drops again.

In security, the mindset couldn't be more different. Everyone wants to get vulnerabilities to zero. But that's an impossible task. You have to pick some points and pick some controls, and see where you can be most effective. That can be a much more efficient strategy of

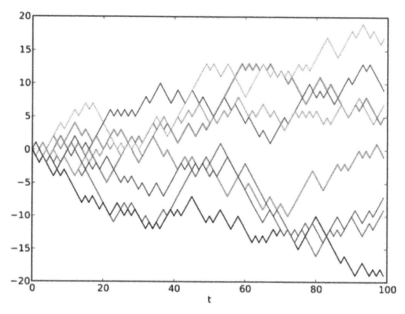

Figure 3.1 Two-dimensional random walk. (© 2013. Reprinted under a CC0 1.0 Universal license [5].)

getting attackers out than just driving down vulnerabilities. One strategy could be capping, or setting a threshold above which you apply swift downward pressure on increasing risk. For example, let's say an enterprise has 100 Windows machines and 100 Linux machines. A vulnerability manager may want to let the number of vulnerabilities rise and fall on the natural course of patching for the Windows machines, unless one crosses the threshold of a 90% chance of exploitation. For the Linux group, patches are less likely to be applied in the natural course of development, so the threshold there could be 70%. This contextual crafting of policies, based on both historical data and what's happening at the enterprise, is the core of the "management" part of vulnerability management.

Markov chains, such as the one in Figure 3.2, are a more sophisticated type of stochastic process. While a random walk has a 50% chance of rising and 50% chance of falling, a Markov chain is a map. You start at state zero, and there's a certain probability of moving to state one and to state two. Some states are interconnected.

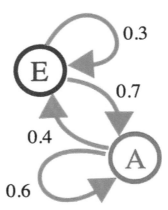

Figure 3.2 Markov chain. (© 2013. Reprinted under a Creative Commons Attribution-Share Alike 3.0 Unported license [6].)

When an exploit is observed in the wild or exploit code is released, there's a certain probability of future actions from attackers. An attacker might see a published vulnerability and write and publish an exploit for it. Other vulnerabilities we don't even know about until we observe someone attacking it in the wild. The attacker found the vulnerability themselves and wrote an exploit for it. Markov chains would allow you to estimate the probability of different attack paths based on those exploits [7].

Similarly, Monte Carlo simulations project many possible outcomes, allowing for better decision-making under uncertainty [8]. In a Monte Carlo simulation, an unknown variable is assigned a set of random values, also known as a probability distribution. The results are calculated over and over, with a different random value within the probability distribution assigned to the unknown variable. In the end, the multiple results are compiled to show the probability of different outcomes.

For example, if you're looking at a CVSS score, it's difficult to know whether you should act on that vulnerability. You lack context about how severe that vulnerability might be to your organization if exploited, and all the potential impacts on revenue, time and effort to remediate, and so on. Monte Carlo simulations allow you to run a number of simulations to determine the likely outcomes and make a more informed decision about whether to prioritize remediating that vulnerability.

In contrast to the stochastic processes above, there's also robust optimization, "a specific and relatively novel methodology for handling optimization problems with uncertain data" [9].

Stochastic processes like random walks, Markov chains, and Monte Carlo simulations all model uncertainties with the assumption that their values are random and obey a known probability distribution. Robust optimization models, on the other hand, "usually contain uncertainties which have distributions that are not precisely known, but are instead confined to some family of distributions based on known distributional properties" [10]. Robust optimization is more oriented toward worst-case scenarios, and can work in conjunction with other stochastic processes depending on the situation.

The central idea here is that in security we have to make decisions under uncertainty—uncertain data, missing data, and a lot under the iceberg when it comes to what we see and what we don't. When optimizing our systems, we need methods that account for that uncertainty, almost as an acknowledgment that our data will never be perfect or certain even in the best-case scenarios. This is largely due to the sentient attacker.

3.1.5 OODA Loops

Colonel John Boyd of the U.S. Air Force developed the observe-orient-decide-act (OODA) loop to help fighter pilots make better decisions in inherently uncertain and dynamic situations. The series of actions is illustrated in Figure 3.3.

OODA loops are a simpler way to think about and manage uncertainty, randomness, and change, with an emphasis on speed. Frans Osinga notes that "In the popularized interpretation, the OODA loop suggests that success in war depends on the ability to out-pace and out-think the opponent, or put differently, on the ability to go through the OODA cycle more rapidly than the opponent" [12, 13].

The pilot who completes the OODA loop fastest will prevail because pilots are responding to situations that have already changed. Using OODA loops to speed decision-making, a pilot might appear unpredictable to the enemy, gaining an advantage.

In vulnerability management, whether it's the speed of an incident response or reacting to new exploits, OODA loops can keep security teams nimble in the face of uncertainty and allow them to

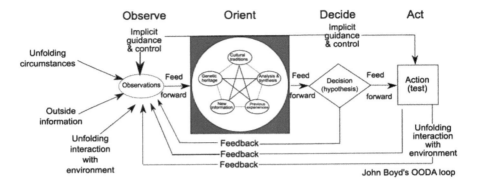

Figure 3.3 Illustration of an OODA loop. (© 2008. Reprinted under the Creative Commons Attribution 3.0 Unported license [11].)

make quick decisions in adversarial situations based on the snapshot of information they have.

The faster and more effectively you can respond to an attack or a newly discovered exploit, the better your organization will be able to stay secure as well as throw off attackers who might perceive your actions as unpredictable compared to those of other organizations.

3.2 MACHINE LEARNING FOR CYBERSECURITY

Using the many sources of data we have and the mathematical modeling we use to compensate for mathematical scale, randomness, and uncertainty, *machine learning* (ML) offers a quick and efficient way to better prioritize decisions and remediate the vulnerabilities that pose the greatest risk.

ML is a classification of artificial intelligence in which algorithms automatically improve over time by observing patterns in data and applying those patterns to subsequent actions. ML allows us to better predict the risk of vulnerabilities and exploits, and continually update recommendations in real time as situations evolve and new threats arise. ML algorithms can adapt based on patterns and changes in the data they analyze, allowing you to create a self-updating system that evolves with the threat landscape as well as your organization.

We can also use *principal component analysis* (PCA) for more efficient data analysis and the creation of predictive models by reducing the dimensionality of multivariate datasets to aid our interpretation of them. While PCA does sacrifice some accuracy, it preserves as much of the dataset's information as possible in representative variables—the principal components—that are easier to analyze and interpret.

The process for creating an ML model is either supervised or unsupervised, but both have advantages in different situations.

3.2.1 Supervised Models

Supervised models essentially learn by example. Using labeled data, both the inputs and outputs are made clear so the algorithms learn how the inputs relate to the outputs. Once the relationship is understood, the ML algorithms can predict outcomes based on new inputs. Two common applications of supervised models are classification and regression. The relationships among types of algorithms are depicted in Figure 3.4.

You can use supervised ML models to estimate the probability that a vulnerability will be exploited, for example. You'll need data on high- and low-risk vulnerabilities, as well as outcome measures such as published exploit (proof of concept) code or real-world exploits.

A quick word on outcome measures: The outcome measure you choose will have an outsized impact on the effectiveness of your model. Consider the use of published exploits versus real-world exploit data. Between 2009 and 2018, there were 9,700 published exploits and 4,200 observed exploits in the wild. Out of all the CVEs (76,000),

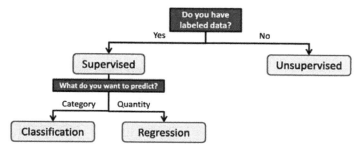

Figure 3.4 Differences among types of algorithms. (© 2016. Reprinted with permission of ICME [14].)

far more had published exploit code (12.8%) than were actually exploited in the wild (5%). Even more significantly, of the vulnerabilities exploited in the wild, only half had associated published code [15]. All of which makes exploits in the wild a more accurate outcome measure to work with.

When thinking about algorithms for supervised models, it's important to remember that each algorithm is most useful at solving a particular type of problem. Here are three of the more common and widely used algorithms you'll need:

- *Regressions.* Regression algorithms take the outcome measure, add several variables, and find a coefficient for each one. The variables that correlate with the outcome measure will allow the algorithm to then predict unknown outcomes based on known parameters. Figures 3.5 and 3.6 depict graphs of linear regressions.
- *Random forests.* A random forest is a collection of decision trees, as illustrated in Figure 3.7. Each tree starts with the outcome measure and branches to the variable that's most likely to contribute to the outcome, and continues to branch subsets. Each tree is randomly created from a selection of features in the larger dataset. Whereas a single decision tree places emphasis on particular features, a random forest chooses features at random, and therefore arrives at more accurate predictions.
- *Neural nets.* This is essentially a multistage way of generating coefficients with some dummy variables. Each stage, or layer, has nodes where a calculation takes place. Each node has a particular weight that's initially randomly assigned and calibrated using the training dataset. If the calculation at a particular node crosses a threshold, it moves the data to the next layer. The result, shown in Figure 3.8, is a neural net that can make a prediction based on the interrelated elements that are most indicative of the outcome [16, 17].

3.2.2 Unsupervised Models

While supervised models offer labeled data and what amounts to an answer sheet for the relationship between inputs and outputs, unsu-

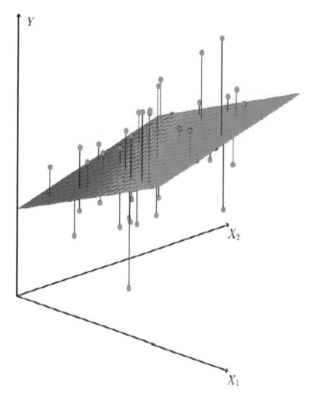

Figure 3.5 Graph of multiple linear regression. (© 2016. Reprinted with permission of ICME [14].)

pervised models free the algorithm to seek out patterns on its own. Figure 3.9 shows the various types of unsupervised models.

Unsupervised models are considered less accurate than supervised models because there is no expected outcome measure to determine accuracy. But that also speaks to its strength and application. Unsupervised models work best when you don't have data on desired outcomes, whether detecting anomalies or clustering data.

Data clustering differs from the classification that supervised models achieve. In classification, ML algorithms rely on labeled data to separate different elements by their features. Clustering, on the other hand, occurs as ML organizes unlabeled features based on their similarities and differences.

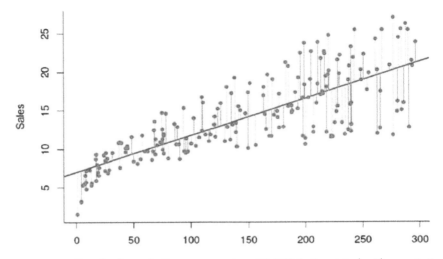

Figure 3.6 Graph of simple linear regression. (© 2016. Reprinted with permission of ICME [14].)

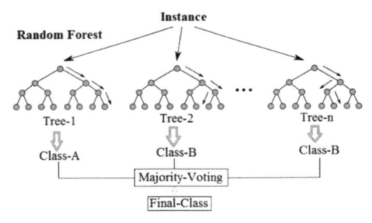

Figure 3.7 Simplified representation of a random forest. (© 2017. Reprinted under the Creative Commons Attribution-Share Alike 4.0 International license [16].)

For example, you could use clustering to examine a pool of malware, determine common characteristics and differences, and group them into different families. Classification, however, could determine whether traffic hitting a server is malicious or not. It's a binary yes-no

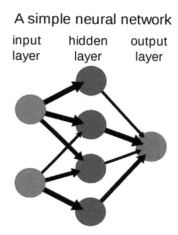

Figure 3.8 Stages of a neural net [20].

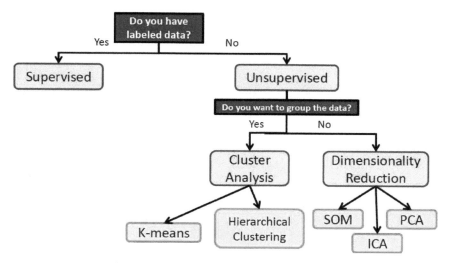

Figure 3.9 Types of unsupervised models. (© 2016. Reprinted with permission of ICME [16].)

choice based on existing knowledge of the characteristics of malicious traffic.

There are two algorithms you should keep in mind for unsupervised learning:

Clustering for malware classification. A common strategy is k-means clustering, which groups data points into k clusters based on common features, with the mean of the data representing the center of each cluster, known as the centroid vector. Figure 3.10 shows a few examples.

For example, one study looked at two datasets of Android malware and used a k-means clustering algorithm to examine and sort their features into three clusters: ransomware, scareware, and goodware.

But one of the most advanced uses of machine learning in security today is malware clustering. In this exercise, you gather machine-level data, such as what the kernel is processing, what the CPU is processing, the NetFlow data, and so on. Once the data is all in one place, you run a clustering algorithm to determine if any of it looks anomalous or resembles the behavior of other malware.

Natural language processing (NLP). NLP is most useful for our purposes in supporting supervised learning ML applications. Text data, such as references to vulnerabilities, blog posts, advisories from Microsoft, and so on can help us create new variables to examine with supervised learning. If you apply a natural language processing algorithm to the available text descriptions of vulnerabilities, you can find similarities or common features.

For example, consider the class of vulnerabilities known as arbitrary remote code execution. There are a couple of ways to figure out that a vulnerability is arbitrary remote code execution. First, you could read the contents of the vulnerability and find that that is what it can cause. The other way is to read all of the text data you have

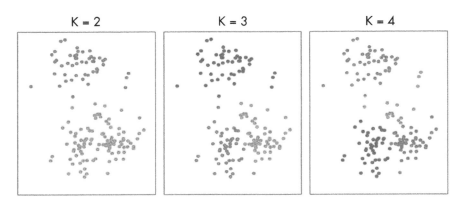

Figure 3.10 Three examples of k-means clustering. (© 2016. Reprinted with permission of ICME [16].)

about that vulnerability and see if the words "arbitrary code" and "remote execution" appear in that text.

An unsupervised algorithm can be applied to vulnerability data—the 160,000 vulnerabilities in NVD, CVSS scores, and their descriptions—to determine the attributes common to arbitrary remote code execution vulnerabilities. We then have to evaluate how predictive that tag is of the outcome we're looking for. That's where supervised learning can step in. NLP operates as a way to extract more features and create more attributes about vulnerabilities that we can use in supervised learning.

Throughout this chapter we've examined the historical and ongoing challenges of risk management and how they've become particularly acute in the age of the internet. Fortunately, we have mathematical models from statistics to game theory to stochastic processes and OODA loops that can help us better understand and manage the risk we face. Machine learning helps us match the scale and complexity of the risk and does so by analyzing data about vulnerabilities and threats, as well as discovering and categorizing the many assets in your organization. All of this will be essential in the next chapter, which is focused on building a decision engine.

References

[1] Bibighaus, D. L., "How Power-Laws Re-Write the Rules of Cyber Warfare," *Journal of Strategic Security*, Vol. 8, No. 4, Winter 2015, https://scholarcommons.usf.edu/cgi/viewcontent.cgi?article=1439&context=jss.

[2] Cook, S. A., "The Complexity of Theorem-Proving Procedures," *Proceedings of the Third Annual ACM Symposium on Theory of Computing*, May 1971, pp. 151–158, https://doi.org/10.1145/800157.805047.

[3] Navarro, D., "Learning Statistics with R: A Tutorial for Psychology Students and Other Beginners," Chapter 10 (translation E. Kothe), January 11, 2019, https://learningstatisticswithr.com/book/estimation.html.

[4] CDC, "Lesson 1: Introduction to Epidemiology," May 18, 2012, https://www.cdc.gov/csels/dsepd/ss1978/lesson1/section1.html.

[5] Németh, L., "Random Walk in Two Dimensions," 2013, https://en.wikipedia.org/wiki/Random_walk#/media/File:Random_walk_2500.svg.

[6] Joxemai4, "Graph of a Markov Chain," 2013, https://commons.wikimedia.org/wiki/File:Markovkate_01.svg#/media/File:Markovkate_01.svg.

[7] Ngoc, L., and D. B. Hoang, "Security threat Probability Computation Using Markov Chain and Common Vulnerability Scoring System," *The 28th International Telecommunication Networks and Application Conference*, November 2018, https://www.researchgate.net/publication/329183104_Se-

curity_threat_probability_computation_using_Markov_Chain_and_Common_Vulnerability_Scoring_System.

[8] Kenton, W., "Monte Carlo Simulation: History, How It Works, and 4 Key Steps," *Investopedia*, August 11, 2022, https://www.investopedia.com/terms/m/montecarlosimulation.asp.

[9] Ben-Tal, A., L. El Ghaoui, and A. Nemirovski, *Robust Optimization*, Princeton, NJ: Princeton University Press, 2009, https://www2.isye.gatech.edu/~nemirovs/FullBookDec11.pdf.

[10] Goh, J., and M. Sim, "Users' Guide to ROME: Robust Optimization Made Easy," September 17, 2009, https://robustopt.com/references/ROME_Guide_1.0.pdf.

[11] Moran, P. E., "Diagram of the OODA Loop," 2008, https://en.wikipedia.org/wiki/OODA_loop#/media/File:OODA.Boyd.svg.

[12] Osinga, F., *Science, Strategy and War: The Strategic Theory of John Boyd*, Delft, The Netherlands, Eburon Academic Publishers, 2005, http://www.projectwhitehorse.com/pdfs/ScienceStrategyWar_Osinga.pdf.

[13] Zager, R., and J. Zager, "OODA Loops in Cyberspace: A New Cyber-Defense Model," *Small Wars Journal*, October 21, 2017, https://smallwarsjournal.com/jrnl/art/ooda-loops-cyberspace-new-cyber-defense-model.

[14] Ioannidis, A., and G. Maher, Fundamentals of Machine Learning Workshop, Institute for Computational and Mathematical Engineering, Stanford University. 2016.

[15] Jacobs, J., S. Romanosky, I. Adjerid, and W. Baker, "Improving Vulnerability Remediation Through Better Exploit Prediction," WEIS 2019, https://weis2019.econinfosec.org/wp-content/uploads/sites/6/2019/05/WEIS_2019_paper_53.pdf.

[16] Maher, G., and A. Ioannidis, *Introduction to Unsupervised Learning*, Institute for Computational and Mathematical Engineering, Stanford University, 2016.

[17] Jagannath, V., "Diagram of a Random Decision Forest," 2017, https://en.wikipedia.org/wiki/Random_forest#/media/File:Random_forest_diagram_complete.png.

[18] Hardesty, L., "Explained: Neural Networks," *MIT News*, April 14, 2017, https://news.mit.edu/2017/explained-neural-networks-deep-learning-0414.

[19] Deep AI, "What Is a Neural Network?" DeepAI.org. https://deepai.org/machine-learning-glossary-and-terms/neural-network.

[20] Wiso, "Neural Network Example," 2008, https://commons.wikimedia.org/wiki/File:Neural_network_example.svg#/media/File:Neural_network_example.svg.

4

HOW TO BUILD A DECISION ENGINE TO FORECAST RISK

The mathematical principles and models behind risk-based vulnerability management serve as the foundation for building a decision engine. That's the aim of this chapter, which will first walk you through the various sources and types of data you'll need before building a logistic regression model and then using neural network models to evaluate different classes of data.

The resulting system can score risk and forecast which vulnerabilities are the most critical to fix. These are hotly debated concepts with as many interpretations as there are risk modelers and forecasters, but for the sake of running a vulnerability management program there are two things we must get good at:

1. Scoring risk is an engine, not a camera [1]. The model does not just measure, but shapes the actions of others. So to build a good risk score is to build a model that will drive the resource and desire constrained actions of those remediating vulnerabilities—IT operations, system administrators, business owners, developers—in the direction of lower risk.

2. Forecasting vulnerability criticality is a continuous improvement process that makes claims about an uncertain future. To do this most usefully to an enterprise, if not necessarily most

accurately, we must take into account the harsh realities of security.

These realities often bump up against the philosophy of machine learning. It is not clear at all that a system as complex as crime perpetrated by a sentient attacker on a self-organizing network of machines in various states of disarray lends itself to forecasts in the first place. So, we must narrow the scope from the Platonic to the useful.

First, we are working with uncertain data and must never assume data covers the range of what is happening or will happen. Second, by the time you've issued a prediction, it is already stale—the data is generated by humans and includes various delays and messiness. And last, making claims about the future is a messy business—we can claim that summer will be hotter than winter, but it is very hard to predict how hot a specific day this summer will be. Robustness with respect to uncertainty, a bias toward precision in modeling, and a nod to our limitations will keep us on course.

4.1 THE DATA

To accurately assess the risk of various vulnerabilities, we need data on the vulnerabilities themselves, diverse threat intel, and an understanding of the assets in your environment. Within those categories, there's much to consider.

First, remember that all of this data is constantly in flux, changing the second you take a snapshot to analyze. New vulnerabilities are discovered tens of thousands of times each year; the assets in your environment come, go, and evolve, and new threats emerge daily. But you can still work from a snapshot to identify and prioritize vulnerabilities.

Different data sources also vary in quality. For example, the descriptions of vulnerabilities can range from thorough and comprehensive to paltry because there is no guidance or style guide for vulnerability descriptions. There's a real problem in security in that some of the best researchers won't write up their vulnerability, and that the most impactful vulnerabilities actually come from attackers who have no incentive to write up a description. As a result of these variations in quality, many algorithms will find vulnerabilities as false positives.

Data is also delivered in various formats. The National Vulnerability Database imposed some structure around reporting, but for the most part, the sources of data are still fragmented. Microsoft issues its own data. Oracle issues its own data. Google will publish a vulnerability to its own bug tracker, but also reserves a CVE in the National Vulnerability Database. We are in the early days of structured security, so different processes and standards are to be expected. We must build a forecast from those disparate sources, which can be as difficult as predicting the weather if every city had its own different weather sensors. This is precisely what was happening across the United States before 1870 when an act of Congress authorized the National Weather Service. Weather was hugely important, but largely a local practice with little hope of accurate forecasts.

All of which makes validating and synthesizing the various sources of information all the more important, a topic we'll cover in this section as well. Ideally, we would have a source of data that's consistent in quality and format the way car repair manuals are. Automotive repair manuals, for example, are written by the same organization with the same objective using the same classifications year after year. While we don't have that in security, we can create our own in a sense by synthesizing various data points into an overall forecast of risk.

Your strategy in approaching the massive amounts of data available is key. Simply collecting as much data as possible and detecting as many vulnerabilities as possible can be less effective because assessing and managing risk is not simply uncovering the signal in the noise. Instead, you should start with the vulnerabilities—the tip of the spear—and determine which ones pose the greatest risk. Then figure out how to identify them in your environment. Data is a lens for the world, and is inherently biased, just as any individual's vision is. By establishing an objective measure the data supports, you can achieve better results. At the very least, your organization can align on the ground truth.

Data is valuable not only as a way to identify potential sources of risk but to put vulnerabilities in context. As we've seen, it's impossible to remediate every vulnerability, but by understanding the assets you have, the vulnerabilities they contain, and the threats you face, you can better prioritize vulnerabilities by the risk they pose to your organization.

4.1.1 Definitions vs Instances

First, we need to distinguish the concepts of definitions and instances. Each is essential, but also gives us different information to work with.

A definition is the description of a vulnerability—what the vulnerability is, where it's found, and what it allows an attacker to potentially exploit. The National Vulnerability Database is full of definitions of thousands of CVEs. When you look up a particular CVE, the description defines a particular flaw. Let's take a look at CVE-2021-21220, a vulnerability in Google Chrome:

```
"cve" : {
    "data_type" : "CVE",
    "data_format" : "MITRE",
    "data_version" : "4.0",
    "CVE_data_meta" : {
      "ID" : "CVE-2021-21220",
      "ASSIGNER" : "chrome-cve-admin@google.com"
    },
    "problemtype" : {
      "problemtype_data" : [ {
        "description" : [ {
          "lang" : "en",
          "value" : "CWE-119"
        }, {
          "lang" : "en",
          "value" : "CWE-20"
        } ]
      } ]
    },
    "references" : {
      "reference_data" : [ {
        "url" : "https://crbug.com/1196683",
        "name" : "https://crbug.com/1196683",
        "refsource" : "MISC",
        "tags" : [ "Third Party Advisory" ]
      }, {
        "url" : "https://chromereleases.google
           blog.com/2021/04/stable-channel-update-
           for-desktop.html",
        "name" : "https://chromereleases.google
           blog.com/2021/04/stable-channel-update-
           for-desktop.html",
        "refsource" : "MISC",
```

```
              "tags" : [ "Release Notes", "Third Party
              Advisory" ]
            }, {
              "url" : "https://security.gentoo.org/
              glsa/202104-08",
              "name" : "GLSA-202104-08",
              "refsource" : "GENTOO",
              "tags" : [ "Third Party Advisory" ]
            }, {
              "url" : "http://packetstormsecurity.com/
              files/162437/Google-Chrome-XOR-Typer-
              Out-Of-Bounds-Access-Remote-Code-Execu
              tion.html",
              "name" : "http://packetstormsecurity.com/
              files/162437/Google-Chrome-XOR-Typer-
              Out-Of-Bounds-Access-Remote-Code-
              Execution.html",
              "refsource" : "MISC",
              "tags" : [ "Third Party Advisory", "VDB
              Entry" ]
            }, {
              "url" : "https://lists.fedoraproject.org/
              archives/list/package-
              announce@lists.fedoraproject.org/
              message/
              VUZBGKGVZADNA3I24NVG7HAYYUTOSN5A/",
              "name" : "FEDORA-2021-c3754414e7",
              "refsource" : "FEDORA",
              "tags" : [ "Mailing List", "Third Party
              Advisory" ]
            }, {
              "url" : "https://lists.fedoraproject.org/
              archives/list/package-
              announce@lists.fedoraproject.org/
              message/
              EAJ42L4JFPBJATCZ7MOZQTUDGV4OEHHG/",
              "name" : "FEDORA-2021-ff893e12c5",
              "refsource" : "FEDORA",
              "tags" : [ "Mailing List", "Third Party
              Advisory" ]
            }, {
              "url" : "https://lists.fedoraproject.org/
              archives/list/package-
              announce@lists.fedoraproject.org/
```

```
            message/
            U3GZ42MYPGD35V652ZPVPYYS7A7LVXVY/",
          "name" : "FEDORA-2021-35d2bb4627",
          "refsource" : "FEDORA",
          "tags" : [ "Mailing List", "Third Party
            Advisory" ]
        } ]
      },
      "description" : {
        "description_data" : [ {
          "lang" : "en",
          "value" : "Insufficient validation
            of untrusted input in V8 in Google
            Chrome prior to 89.0.4389.128 allowed a
            remote attacker to potentially exploit
            heap corruption via a crafted HTML
            page."
        } ]
      }
    },
    "configurations" : {
      "CVE_data_version" : "4.0",
      "nodes" : [ {
        "operator" : "OR",
        "children" : [ ],
        "cpe_match" : [ {
          "vulnerable" : true,
          "cpe23Uri" : "cpe:2.3:a:google:
            chrome:*:*:*:*:*:*:*:*",
          "versionEndExcluding" : "89.0.4389.128",
          "cpe_name" : [ ]
        } ]
      }, {
        "operator" : "OR",
        "children" : [ ],
        "cpe_match" : [ {
          "vulnerable" : true,
          "cpe23Uri" : "cpe:2.3:o:fedoraproject:
            fedora:32:*:*:*:*:*:*:*",
          "cpe_name" : [ ]
        }, {
          "vulnerable" : true,
          "cpe23Uri" : "cpe:2.3:o:fedoraproject:
            fedora:33:*:*:*:*:*:*:*",
          "cpe_name" : [ ]
```

4.1 THE DATA

```
      }, {
        "vulnerable" : true,
        "cpe23Uri" : "cpe:2.3:o:fedoraproject:
         fedora:34:*:*:*:*:*:*:*",
        "cpe_name" : [ ]
      } ]
    } ]
  },
  "impact" : {
    "baseMetricV3" : {
      "cvssV3" : {
        "version" : "3.1",
        "vectorString" : "CVSS:3.1/AV:N/AC:L/
         PR:N/UI:R/S:U/C:H/I:H/A:H",
        "attackVector" : "NETWORK",
        "attackComplexity" : "LOW",
        "privilegesRequired" : "NONE",
        "userInteraction" : "REQUIRED",
        "scope" : "UNCHANGED",
        "confidentialityImpact" : "HIGH",
        "integrityImpact" : "HIGH",
        "availabilityImpact" : "HIGH",
        "baseScore" : 8.8,
        "baseSeverity" : "HIGH"
      },
      "exploitabilityScore" : 2.8,
      "impactScore" : 5.9
    },
    "baseMetricV2" : {
      "cvssV2" : {
        "version" : "2.0",
        "vectorString" : "AV:N/AC:M/
         Au:N/C:P/I:P/A:P",
        "accessVector" : "NETWORK",
        "accessComplexity" : "MEDIUM",
        "authentication" : "NONE",
        "confidentialityImpact" : "PARTIAL",
        "integrityImpact" : "PARTIAL",
        "availabilityImpact" : "PARTIAL",
        "baseScore" : 6.8
      },
      "severity" : "MEDIUM",
      "exploitabilityScore" : 8.6,
      "impactScore" : 6.4,
      "acInsufInfo" : false,
```

```
            "obtainAllPrivilege" : false,
            "obtainUserPrivilege" : false,
            "obtainOtherPrivilege" : false,
            "userInteractionRequired" : true
        }
    },
    "publishedDate" : "2021-04-26T17:15Z",
    "lastModifiedDate" : "2021-06-01T15:20Z"
}
```

This is a definition. It tells you about the flaw, which software it affects, how it works from a technical perspective, the impact if it is taken advantage of, and all of the associated metadata. More on this later.

```
CVE-2020-1147
Tenable.io.via.xml... Tenable Nessus XML Port 445
A remote code execution vulnerability exists in
.NET Framework, Microsoft SharePoint, and Visual
Studio when the software fails to check the
source markup of XML file input, aka '.NET
Framework, SharePoint Server, and Visual Studio
Remote Code Execution Vulnerability'.

CVSS 2: 7
CVSS 3: 7.8
Scanner IDs
138460

Unique Identifiers
138460

Asset HST-38cc
IP ADDRESS
10.15.15.162
HOSTNAME
10-15-15-162.us-east.randomhost.com
NETBIOS
HST-38cc
OPERATING SYSTEM
Microsoft Windows Server 2008 R2","Microsoft
Windows XP Professional","Microsoft Windows 10
TYPE
OPEN PORTS
```

```
445   (TCP)
3389  (TCP)
49161 (TCP)
```

Now we're looking at a definition, namely CVE-2020-1147, on a particular machine, particularly the Windows machine that has IP 10.15.15.162 and was scanned by a Tenable.io scanner, which discovered this vulnerability. This is an instance. An organization with 2,000 Windows machines may have 2,000 instances of CVE-2020-1147 across the enterprise. One definition, one CVE, thousands of machines to protect.

An instance is a vulnerability in context that helps you decide whether to escalate or de-escalate that risk. That context could be what machine the vulnerability is running on or what organization that machine is part of. For example, a patent data server running Windows 7 at Lockheed Martin is a much more vulnerable device than Windows 7 on a home computer.

In addition, Windows 7 doesn't run by itself; you might have a firewall on it, you might have another application that allows another vulnerability to jump to that machine.

Definitional analysis is useful in vulnerability management, but to make an informed decision about whether to remediate a particular vulnerability, you need many more contextual sources about it. That's what security practitioners who have adopted a *modern vulnerability management* (MVM) practice do. For example, an enterprise organization might have decided that 2% of its vulnerabilities are risky and need to be fixed. At an organization of that size, 2% could easily be 200,000 vulnerabilities. The firm will still need to uncover which vulnerabilities pose the most risk, and they do so by finding the context that escalates or de-escalates remediation for each one.

4.1.2 Vulnerability Data

4.1.2.1 *Vulnerability Assessment*

There are three main ways to assess vulnerabilities in your environment: network scans, authenticated scans, and asset inventory. Each has its strengths and weaknesses, and the strongest programs employ a variety of methods for assessing vulnerabilities to ensure a well-rounded, comprehensive approach.

A network scan uses IP addresses to ping the machines on a network, and receives back data about those assets. That data includes details like what operating system the asset is running, and what function it performs—is it an end-user device or a database? Using the data received from the device, you can determine what vulnerabilities might be present and how critical the asset is to your organization. Network scans are automated and you don't need to install anything to perform one. But while they're fast and easy, they're more likely to produce false positives.

An authenticated scan goes a step further. You can collect even more information about what's running on each asset via an insider's view. Because the scan is authenticated, it has privileged access to the assets. You can set up a botnet with a command and control server that sends commands and receives responses from each asset. The authenticated scanner checks the kernel for what applications the asset is running, what files have been uploaded, and so on. It's a much more precise way to identify vulnerabilities on a machine. The downside is that authenticated scans take up more resources.

For example, if you're running an agent that's constantly scanning for vulnerabilities, you can encounter unintended consequences. On a machine that's processing credit card payments and is already at 99% capacity, the scanning agent could cause it to skip a payment or two. In that situation, the scanning for vulnerabilities had a negative impact on the business. This was a major problem in the 2000s that has been mitigated somewhat by the cloud, but still exists and should be considered when using authenticated scans.

The third way is an asset inventory, which tracks machines and applications they're running. For example, your sales team is given the same laptop, and you know they come preinstalled with Adobe Reader 9.1. You know that Adobe Reader 9.1 is affected by seven vulnerabilities, so you can safely assume that every laptop you've given to the sales team is affected by those vulnerabilities.

Each method has blind spots and could miss assets or the applications or files on those assets. Each will produce false positives. Ultimately you'll need all three to get a comprehensive view of vulnerabilities in your environment. We've seen Fortune 100 companies use multiple vendors for all three methods, using a specific type of scanner for their credit card processing, for example, one they know is operationally safe.

4.1.2.2 Static Application Security Testing and Dynamic Application Security Testing

In the previous section, all of the vulnerability scanning we've talked about is based on CVE descriptions. Each CVE consists of two parts: a CWE and a CPE.

A CWE is a type of flaw, such as a man-in-the-middle attack, buffer overflow attack, improper credential authentication, or a password in plain text.

```
https://cwe.mitre.org/data/definitions/79.html
CWE-79: Improper Neutralization of Input During Web
Page Generation ('Cross-site Scripting')
Weakness ID: 79
Abstraction: Base
Structure: Simple
Status: Stable
Presentation Filter:
Complete
+ Description
The software does not neutralize or incorrectly
neutralizes user-controllable input before it is
placed in output that is used as a web page that is
served to other users.
+ Extended Description
Cross-site scripting (XSS) vulnerabilities occur
when:

Untrusted data enters a web application, typically
from a web request.

The web application dynamically generates a web
page that contains this untrusted data.
During page generation, the application does not
prevent the data from containing content that is
executable by a web browser, such as JavaScript,
HTML tags, HTML attributes, mouse events, Flash,
ActiveX, etc.
```

```
A victim visits the generated web page through a
web browser, which contains malicious script that
was injected using the untrusted data.

Since the script comes from a web page that was
sent by the web server, the victim's web browser
executes the malicious script in the context of the
web server's domain.

This effectively violates the intention of the web
browser's same-origin policy, which states that
scripts in one domain should not be able to access
resources or run code in a different domain.
```

A CPE shows where that flaw resides. It gets very specific. For example, a CPE might talk about improper credential handling in Microsoft Word 2016 Service Pack 1 version 1.025.3 [2].

```
cpe:2.3:o:microsoft:windows_10:-:*:*:*:*:*:*:*
cpe:2.3:o:microsoft:windows_server_2008:r2:sp1:*:*:
*:*:itanium:*
```

Combined, they show the type of bug on the type of software. This is the innovation that is the National Vulnerability Database. But the NVD could never document all of the bugs in all of the software in the world because a massive amount of software is designed to perform internal processes and services inside businesses. They're built and used internally, often by a single company, and they also have bugs and vulnerabilities.

Static application security testing (SAST) and dynamic application security testing (DAST) are ways to find those vulnerabilities in code that has no integrations with a common platform the way those in the NVD do.

Insurance company Allstate has about 1,400 internal applications that its employees have written over the years, from internal billing and portal management to insurance signup and assessment. Kenna Security has six applications that it maintains—we have one application that manages the natural language processing and one that predicts which vulnerabilities are most likely to be exploited.

Neither of those are going to be included in the NVD. But the coders of those applications still make mistakes all the time.

SAST examines those weakness enumerations without a platform integration. There are a number of vendors that examine the status of code based on a checklist. They plug into your code, read your GitHub repository, read your machine Python code, and then point out where you didn't sanitize the input, for example, and indicate that it is probably going to create a vulnerability. During the development life cycle, the code will pass through these checks and produce a number of vulnerabilities for which we have no data. All we know is this vulnerability exists in this one application.

The good news is that it's probably less of a target for attackers. If you write an exploit for one of these vulnerabilities, there's not going to be a huge payoff because there's only one company that's running this software. The bad news is that it's much easier to exploit.

DAST is the network analog to SAST, done by white hat security. They log on to a company's website and run a number of commands that indicate the types of vulnerabilities present. Burp Suite is one open-source DAST tool. Point it at a web application and it will run various checks and then tell you that you have cross-site scripting on a form, or if an attacker inputs a particular numeric character in a particular form, they can bypass security. Some of the weaknesses are more exploited than others—*Structured Query Language* (SQL) injections being a common example.

CVEs are application security findings in specific applications. But the world is full of software that is never inventoried or logged, and that's where SAST and DAST play a role. Ultimately all describe weaknesses in software, whether named or not.

CVE enumeration is key to the machine learning work we'll describe in more detail later. At their core, all vulnerabilities are write-ups of flaws in the form of large text documents. CVEs offer a common identifier, an index to the documents that allows security practitioners to expand on the specifics of the flaw.

In application security, when SAST and DAST scanners don't have a CVE, they produce findings specific to the code the scan is run on. The more specific the document, the less information it's likely to contain. CWE attempts to remedy this issue by creating an identifier that works across different findings. Unfortunately, it's convoluted at best.

CWEs exist in a nested hierarchy of parent and child relationships and those relationships and the subsets of the CWEs shift depending on which of the various views you're using to look at them. Different views all work with a subset of the core enumerations and implement varying relationships among the enumerations. If you select any single view, you won't have a map across views and possible data feeds.

What's needed for better analysis of application security vulnerabilities is a taxonomy derived from data. This is probably best done by a vendor that has access to a large number of appsec vulnerability findings. We would suggest using natural language processing to update and refine the CWE mappings.

Application security vulnerabilities are harder to remediate, but there are fewer of them. Prioritization has less value in the face of less voluminous data, but triaging the vulnerabilities is still of value. Synopsys has done some good work exploring how machine learning can automate the process of triaging application security vulnerabilities [3].

4.1.3 Threat Intel Sources

To power a MVM platform, you need a certain breadth and depth of contextual threat and vulnerability intelligence so that your predictive risk-scoring algorithms are as precise and accurate as possible. You do this through threat and exploit intelligence feeds.

You need many feeds to cover all of the threat and vulnerability data categories. A small number of feeds leaves you with less than stellar coverage. It's easy to see why when considering the many different threat and vulnerability data categories (Figure 4.1).

A threat feed is all the information we collect from a single source. For example, Reversing Labs and Exodus Intelligence are both feeds. A threat or vulnerability category is exactly what it sounds like: a category of threat and/or vulnerability data. Each feed can provide data on one to six threat or vulnerability categories, such as the Chatter and Exploit Databases, but more commonly, a feed will only supply data for one category. To achieve comprehensive coverage, you need multiple feeds. Without an adequate number, you won't have enough coverage to provide high-fidelity risk prioritization throughout the CVE life cycle.

Threat and vulnerability categories vary in their relative importance in predicting vulnerability risk and in their utility in predicting risk at different points in the CVE life cycle—CVE Named,

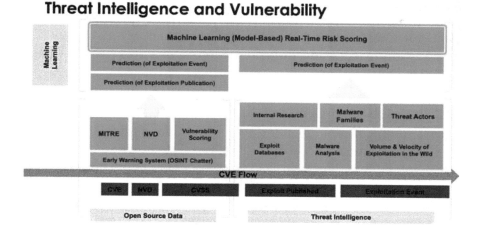

Figure 4.1 Threat intelligence and vulnerability categories. (© 2020 Kenna Security. Reprinted with permission [4].)

CVSS Score Assigned, Exploit Released, and Exploitation in the Wild (Figure 4.1). For example, the category "chatter" is extremely important in helping to score a vulnerability in the early stages of the CVE life cycle and becomes less important as other categories of threat and vulnerability data become available.

To get the best possible risk prioritization at every stage of the CVE life cycle you need to cover all the threat or vulnerability categories we have listed. But even if you have six feeds and cover all of the threat vulnerability categories, that full coverage isn't enough. In this scenario, you have breadth but are lacking in depth. To make up for this, we need to make sure each threat or vulnerability category is deeply covered by our feeds as well.

That depth comes from the data science we discussed earlier in this chapter. Using machine learning models, we know which vulnerabilities were successfully exploited, so we can measure the predictive accuracy of our scoring algorithms and contextual data. Using those models, you can test each feed to see if it improves your predictive accuracy. If it doesn't pass muster, don't add it to the mix.

For example, at Kenna Security, we use more than 15 threat intelligence feeds, and our security research team is always looking at and evaluating new feeds. In fact, just last year we added a feed that bolstered our risk prediction capabilities in the early stages of the

CVE life cycle with a wealth of pre-NVD chatter information (certain types of chatter occur before a CVE is published to the NIST NVD database). Our research team consistently partners with our feed providers to improve the breadth and depth of category coverage in their data to enhance vulnerability risk scoring.

The various threat intelligence feeds fall into five categories:

1. *Internal intelligence.* A good place to start in gathering threat intelligence is your organization's own assets and behavior. Gather and analyze data from firewall logs, DNS logs, network event logs. Examine reports of past incident responses.

2. *Network intelligence.* Analyze traffic at your organization's network boundary and on external networks by examining network data packets. Learn who is accessing your network, how, when, and from where—both physical location and IP address. FireEye is a good example of a network intelligence tool.

3. *Edge intelligence.* Keep an eye on host activity at the edge of your network. Governments, *internet service providers* (ISPs), and telecoms all have useful data to feed analysis. Akamai is an example of an edge intelligence tool.

4. *Open-source intelligence.* Gather information on threats from publicly available sources, such as websites and blogs, social media, message boards, and news feeds.

5. *Closed-source intelligence.* This information is gleaned from dark web forums, intelligence agencies and law enforcement, and human intelligence. One example is the Financial Services Information Sharing and Analysis Center (FS-ISAC), which offers information on security threats to the financial services sector.

4.1.4 Asset Discovery and Categorization: Configuration Management Database

In addition to gathering data on vulnerabilities and threats, you also need as clear a picture as possible of what assets you have as part of your environment. This is one of those areas in security where perfection is the enemy of the good. In fact, it's impossible for most organizations to achieve 100% coverage of their assets.

That's because those assets are always changing, especially as organizations increasingly move to the cloud and turn up and turn down instances all the time. So the assets you know about right now might be different in an hour. That's OK.

Begin with the basics. A lot of organizations have already started down the asset discovery path. Network discovery tools like Nmap discover hosts and services on your network by pinging them and interpreting the responses.

In addition to scanning, some organizations are employing agents on hosts that broadcast back every application and service installed and running on that machine. The discovery from that agent reveals every asset on the network and allows you to determine whether you have problematic security controls. The tools are shifting to be less network-based and more agent-based.

Once you discover these assets, you need to categorize them as internal and external. This is critical to controlling your attack surface since attackers may already know about vulnerabilities or exploits in external assets even if you don't. The next, more difficult step is determining who owns those assets and who to contact should they need patching or fixing.

Whether you load that asset scanning data into a configuration management database (CMDB) or an asset management system, anything you can do to start to get a handle on your assets is critical. All the vulnerability data and threat intelligence in the world can't help you if you don't know what assets you have in your environment that could be vulnerable to those threats.

A CMDB manages configuration data for each asset, or configuration item (CI), defined by the Information Technology Infrastructure Library (ITIL) as any component that needs to be managed to deliver an IT service. That includes information specific to each asset, including:

- Type of CI (hardware, software, etc.);
- Attributes and data, including a unique identifier code;
- Who owns each CI;
- How important it is to the organization;

- Whether an asset contains sensitive data and falls under the payment card industry (PCI), Sarbanes-Oxley Act (SOX), or other compliance rules;
- Relationships among CIs, including dependencies.

This will take some time to define and map out. Some organizations already have CMDBs but the CIs are poorly defined, disorganized, or out of date. Automating updates for your CMDB is essential.

Once you've scanned and categorized your assets, you can correlate your CMDB with your vulnerability data to determine the risk each vulnerability poses. Of course, as we've mentioned already, mapping all your assets and ensuring you have categorized them in your CMDB is a never-ending process. Again, start small and scale. Especially in large organizations, mapping out all the assets in the entire firm will be time consuming and involve complex relationships among assets, as well as other challenges.

What's important is that you start somewhere and have as accurate and current a picture of your environment as possible so you can best assess which vulnerabilities could disrupt your business operations.

4.1.5 Data Validation

We've covered the various types of data that are crucial to measuring risk that inform the models we discussed earlier in this chapter. Once you begin assessing vulnerabilities, collecting threat intel, and mapping out the assets in your organization, you need to ensure you have accurate and quality data. Garbage in, garbage out, as the saying goes.

As we discussed at the start of this section, the sources of data you're pulling from can vary widely in terms of quality and format. It might include duplicates or null values. It could become corrupted when moved due to inconsistencies. Validating your data saves time, effort, and money. So before you import and process data, it's important to validate it using an extract, transform, and load (ETL) process.

4.1.5.1 ETL

You can build your own ETL tool or use one of the well-known tools that will automate the process. Your exact requirements will depend on your organization, but be sure any ETL tool can connect to the data sources you're using and scale with your organization as your data grows. Price and security are two other major criteria to consider.

Different data sources use different formats, including Extensible Markup Language (XML), flat files, relational databases, RSS feeds, and more. In the extraction phase, we export data to a staging area where we can begin to organize it. Depending on the sources, format, and volume of the data, this can be a complex and time-consuming step. Extracting unstructured data in particular can be challenging and may require specialized tools. This first step is an important one to avoid damaging your data warehouse with data that's corrupted or in the wrong format.

From there you need to cleanse the data. The transformation stage involves filtering, deduplicating, and authenticating. This is also when you format the data to match the schema of your data warehouse, applying rules to organize it according to your own systems and bringing it all into alignment for processing.

Loading the data from the staging area to your target data warehouse is the final step. After the initial load, ongoing and incremental loads will keep data up to date.

4.2 BUILDING A LOGISTIC REGRESSION MODEL

The vulnerability scoring model we're setting out to build should be:

- *Simple.* Keep variables to a bare minimum.
- *Explainable.* We should see a consistent and intuitive cause/effect relationship between vulnerability attributes and the rating.
- *Defensible.* As we apply science, we should get feedback and performance measurements from the real-world data.
- *An improvement.* We should see a measurable improvement over an existing solution.

While these are the most important objectives, keep some other attributes in mind as well:

- The target variable is actively exploited vulnerabilities, not published exploit code. This will be highly unbalanced data since roughly 2% of vulnerabilities are exploited in the wild.

- Ratings should be set in terms of a probability instead of a scoring scale, although both can be projected into a 0–100 scale.
- The barrier to run the model in production should be as low as possible.

To achieve these objectives, we'll use a logistic regression algorithm. That decision will immediately make our vulnerability scoring model explainable and defensible as well as relatively simple to execute in any production environment. It will also be able to produce probabilities. We can meet the other two main objectives through feature engineering and measuring the performance of the model.

4.2.1 Data Sources and Feature Engineering

We'll use the following categories of data to build our model:

- CVSS tags;
- CPE vendors;
- CPE products;
- Reference lists—presence on a list and a count of how many references the CVE has;
- Published exploits—yes or no, and which framework the exploit appears in;
- Vulnerability prevalence—a count of how often a vulnerability appears;
- CVE tags—tags drawn from the descriptions of scraped references for each CVE;
- Target variable.

Altogether, there are 291 data points gathered for CVEs published 2009 and later. Among that data, there are 64,536 CVEs, 1,291 of which are labeled as being exploited in the wild.

4.2.1.1 Feature Engineering

There are multiple strategies for feature engineering. For example, we could reduce the feature set in each category, then further reduce those as a whole. While that would have the benefit of speed, it would not yield optimal results.

4.2 BUILDING A LOGISTIC REGRESSION MODEL

One of the better strategies is to apply a stepwise technique, evaluating the benefit each feature would add to the model by measuring the Akaike information criterion (AIC) of the model. With this technique we can manually build up the feature set one variable at a time and record the improvement in AIC (lower AICs are better) (Figure 4.2).

Each variable is added one by one, in the order shown on the plot (top to bottom) based on the influence it has on reducing the AIC. "any_exploits" is the most influential variable with an AIC of 10,545. By adding "vuln_count" (the count of findings for this vulnerability) brought the AIC down to 8,375. Additional variables continue to reduce the AIC down the line. The plot shows how the reduction of AIC starts to flatten out, indicating that adding more variables may result in only a very minor improvement in AIC reduction.

As you may imagine, training stepwise models can be a long process. As a way to cut training time, we can filter out the bottom half of the most prevalent variables. For example, PHP as a product only had 339 CVEs associated with it, so its overall influence wouldn't be that significant. As a result, it can be removed along with 142 other variables.

There is another potential issue to watch out for: Variables that create a probability of 0 (or 1) can artificially inflate (or deflate) the

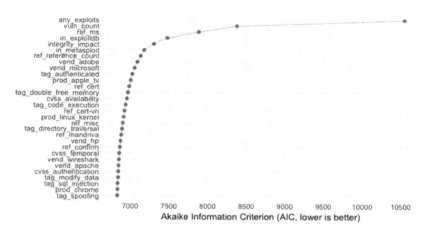

Figure 4.2 Ranking features by AIC. (© 2018 Kenna Security. Reprinted with permission [4].)

output. For example, there is one variable in the above list of top 30 that we manually removed: "prod_apple_tv." There are 589 CVEs since 2009 that list the Apple TV, but none of them were ever recorded to be exploited. For future work, we may be able to reduce the feature list to only those that show a split across vulnerabilities exploited in the wild and those not exploited.

4.2.1.2 Interpretation of Features

By generating a logistic model using the top 29 variables (dropping Apple TV), we can visualize the weights each feature is receiving as well as our confidence around that estimate (see Figure 4.3).

We will go into a deep dive of these values and their meaning later in this chapter. For now, remember that these represent the log odds ratio and they are additive. So if an exploit is published, the log odds are increased by 4.5 and another 0.97 if it's in Metasploit.

That these are log odds does make explainability more difficult, but possible communication methods are explored in the section below. Communicating the model and helping consumers understand the attributes that contribute to higher or lower probabilities should be a major discussion point prior to the rollout into production.

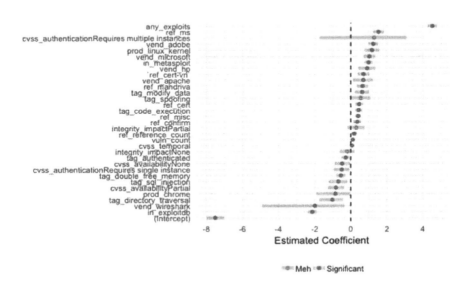

Figure 4.3 Weights of each feature and confidence in that estimate. (© 2018 Kenna Security. Reprinted with permission [4].)

4.2.2 Testing Model Performance

To test the performance of the model, we can use a five-fold cross-validation technique, using four-fifths of the data to train a model and predicting on the remaining fifth. This enables us to produce a prediction for each CVE without using it to train the model that predicted it.

In Figure 4.4, the first view shows the predicted probability between CVEs that have been exploited in the wild versus those that haven't. Exploited CVEs should have a high probability (farther to the right) and CVEs not yet exploited should receive a low probability (farther to the left).

The plot in Figure 4.4 looks fairly good. The majority of CVEs that haven't been exploited (top plot) mostly received low predictions. While the "no" portion looks like nothing scored above 5%, there are a few thousand CVEs spread out above 5%, including 650 unexploited CVEs predicted to be above 20% chance of exploitation. The "yes" portion also has potential. There is obviously a much better spread here and even above 90% probability of exploitation is included.

4.2.2.1 Calibration Plot

A good way to check the predicted probabilities is to create a calibration plot. This plots the prediction (using the probability of ex-

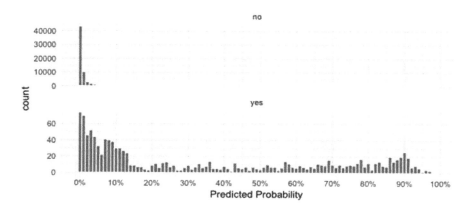

Figure 4.4 Predicted probability between CVEs exploited in the wild versus those that haven't. (© 2018 Kenna Security. Reprinted with permission [4].)

ploitation) against reality (using the percent actually observed to be exploited).

This plot will group the predictions together around a value, for example 10%, and then check how many in that group were actually exploited. In the 10% grouping, we should see about 10% of them being exploited in the wild. The calibration plot in Figure 4.5 has a dashed line that represents perfect calibration and the shaded gray region represents our confidence around the accuracy of the measurement. A good calibration plot should follow the dashed line across the plot.

Overall, this looks quite good. There is a little bit of underestimation from 65% to 85%, and then some overestimation over 90%, but overall this is fairly well calibrated.

4.2.2.2 Simplicity vs Performance

There is a benefit in keeping the model simple by reducing the number of variables, but that often comes at a cost to model performance. We can use the measurements we created in the last section (average improvement over the level of effort) and calculate that for very simple models (with five variables) up to the full list of 29 variables (where the AIC started to level off) (Table 4.1).

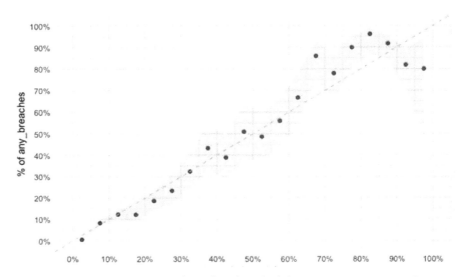

Figure 4.5 Calibration plot of predicted probabilities. (© 2018 Kenna Security. Reprinted with permission [4].)

4.2 BUILDING A LOGISTIC REGRESSION MODEL

Table 4.1
The 29 Variables (in Order) for Reference

1	any_exploits	11	ref_cert	21	ref_confirm
2	vuln_count	12	tag_double_free_memory	22	cvss_temporal
3	ref_ms	13	cvss_availability	23	vend_wireshark
4	in_exploitdb	14	tag_code_execution	24	vend_apache
5	integrity_impact	15	ref_cert-vn	25	cvss_authentication
6	in_metasploit	16	prod_linux_kernel	26	tag_modify_data
7	ref_reference_count	17	ref_misc	27	tag_sql_injection
8	vend_adobe	18	tag_directory_traversal	28	prod_chrome
9	vend_microsoft	19	ref_mandriva	29	tag_spoofing
10	tag_authenticated	20	vend_hp		

You can see the diminishing returns in coverage (the percentage of vulnerabilities correctly identified for remediation) and efficiency (the percentage of vulnerabilities remediated that were exploited) as the number of variables rises (Figure 4.6).

Using this visualization, and depending on how important simplicity is, we can see the trade-off between reducing the number of

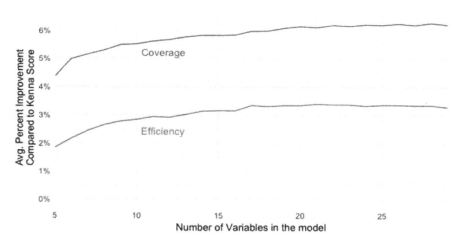

Figure 4.6 Visualization of the trade-off between reducing the number of variables and the average percent improvement in coverage and efficiency. (© 2018 Kenna Security. Reprinted with permission [4].)

variables and the average percent improvement in both the coverage and efficiency.

4.2.3 Implementing in Production

There are several steps to implementing this in production, but there are no special libraries or complex algorithms to support. Logistic regression is a simple additive model. Each variable has an associated weight and that weight is multiplied against the variable and those are summed up to get the log odds from the logistic regression model. But there are a few steps in the overall process to implement in production:

1. Data preparation;
2. Application of the model;
3. Conversion from log odds to probability.

We will walk through each of these steps in turn and then discuss possible methods to communicate the predicted probabilities.

4.2.3.1 Data Preparation

Most of the variables are encoded as 1 if present or 0 if not present. For example, if the CVE has a published exploit, then "any_exploit" will be a 1. Otherwise, it will be a 0. Table 4.2 is a list of the data preparations and encodings.

Note that several of the variables are encoded as "factor of options." Those will be converted into 1 or 0 types of encoding when we talk about the weights associated with these variables.

4.2.3.2 Application of the Model

In addition to encoding all the variables, we must have weights associated with each variable. The weights for each variable are stored in a CSV file [5], which has more degrees of precision, but rounded values are shown below in Table 4.3.

Note that the "factor with options" variables from above are expanded out in Table 4.3. For example "cvss_authentication" now has two variables: "Requires multiple instances" and "Requires single instance." Those will be 1 if the text value of "cvss_authentication" matches the name.

The model is constructed with the following formula:

4.2 BUILDING A LOGISTIC REGRESSION MODEL

Table 4.2
Reference for Binary Variables of the Model

any_exploits	1 if exploit code is published
vuln_count	Natural log of vulnerability count
ref_ms	1 if CVE has reference to MS
in_exploitdb	1 if published exploit appears in ExploitDB
integrity_impact	Factor of options (discussed below)
in_metasploit	1 if published exploit appears in Metasploit
ref_reference_count	Natural log of the count of references in CVE
vend_adobe	1 if CPE vendor is Adobe
vend_microsoft	1 if CPE vendor is Microsoft
tag_authenticated	1 if CVE is tagged with "authenticated"
ref_cert	1 if CVE has reference to "cert" list
tag_double_free_memory	1 if CVE is tagged with "double free memory"
cvss_availability	factor of options (discussed below)
tag_code_execution	1 if CVE is tagged with "code execution"
ref_cert-vn	1 if CVE has reference to "cert-vn" list
prod_linux_kernel	1 if CPE product is "Linux kernel"
ref_misc	1 if CVE has reference to "MISC"
tag_directory_traversal	1 if CVE is tagged with "directory traversal"
ref_mandriva	1 if CVE has reference to "Mandriva"
vend_hp	1 if CPE vendor is HP
ref_confirm	1 if CVE has reference to "confirm"
cvss_temporal	CVSS temporal score, unmodified
vend_wireshark	1 if CPE vendor is "Wireshark"
vend_apache	1 if CPE vendor is "Apache"
cvss_authentication	Factor of options (discussed below)
tag_modify_data	1 if CVE is tagged with "modify data"
tag_sql_injection	1 if CVE is tagged with "SQL injection"
prod_chrome	1 if CPE product is "chrome"
tag_spoofing	1 if CVE is tagged with "spoofing"

$$Log\ Odds = bias + W_1 X_1 + W_2 X_2 + \ldots + W_n X_n \quad (4.1)$$

Table 4.3
Model Coefficients

term	Coef
cvss_temporal	0.020
cvss_authenticationRequires multiple instances	0.137
cvss_authenticationRequires single instance	−0.515
integrity_impactNone	−0.339
integrity_impactPartial	0.092
cvss_availabilityNone	−0.401
cvss_availabilityPartial	−0.711
tag_authenticated	−0.373
tag_code_execution	0.347
tag_directory_traversal	−0.838
tag_double_free_memory	−0.478
tag_modify_data	0.580
tag_spoofing	0.455
tag_sql_injection	−0.624
prod_chrome	−0.579
prod_linux_kernel	0.890
vend_adobe	1.163
vend_apache	0.650
vend_hp	0.758
vend_microsoft	0.843
vend_wireshark	−0.692
ref_cert	0.474
`ref_cert-vn`	0.647
ref_confirm	0.252
ref_mandriva	0.658
ref_misc	0.374
ref_ms	1.435
ref_reference_count	0.116
vuln_count	0.114
any_exploits	4.284
in_metasploit	1.015
in_exploitdb	−1.976
Bias/Intercept	−6.630

4.2 BUILDING A LOGISTIC REGRESSION MODEL

where:

- Log odds is the natural log of the odds ratio;
- Bias is the bias/intercept value in Table 4.4 (−6.630);
- Wi is the *i*th weight in the list above (1..n);
- Xi is the *i*th variable in the list above (1..n).

As an example, suppose a CVE has none of the variables above, it has no published exploit, no vulnerabilities observed, and none of the tags, products, or vendors. All of the encoded variables would be zero, so the formula (4.2) is simply

$$Log\ Odds = bias = -6.63 \tag{4.2}$$

Since all of the variables (Xi) are zero, multiplying by the weights will always be zero.

Let's take a real-life example, CVE-2016-1499 (Table 4.4). If the variable is not shown, it's zero.

The variable encoding from step 1 is in the "Variable" column, the weight from the table above is in the "Weight" column, and those are multiplied together to get the "Product" column. That column is then added together to get the "Log odds" value on the bottom (in this case, −6.8621…). The real value may be slightly different due to rounding issues in the formatting here.

Table 4.4
Model Calculations for CVE-2016-1499

Term	Variable	Weight	Product
Bias/intercept		-6.630	-6.63
cvss_temporal	8	0.020	0.1624637043
cvss_authenticationRequires single instance	1	-0.515	-0.5153246766
integrity_impactNone	1	-0.339	-0.3389920231
tag_authenticated	1	-0.373	-0.3731910107
ref_confirm	1	0.252	0.2519884216
ref_misc	1	0.374	0.3736629382
ref_reference_count	1.7912	0.116	0.2078441
Log odds		(sum)	-6.862177559

Let's also run this for CVE-2016-0016 (Table 4.5).

It's obvious (from a security standpoint) that CVE-2016-0016 should be rated much higher compared to CVE-2016-1499, and the log odds show that.

4.2.3.3 Converting Log Odds to Probability

Log odds can be converted to a probability with this formula:

$$\frac{e^{odds}}{1+e^{odds}} \tag{4.3}$$

where "odds" are the log odds.

To interpret that, if a vulnerability has none of the values in the feature list, the log odds would be equal to the (intercept) value, or -6.63. Plugging in -6.63 into the formula we would predict the probability of that being exploited in the wild to be 0.0013 or 0.13%.

Looking at CVE-2016-1499 from the previous section:

$$\frac{0.00104^{e-6.862=0.00104}}{1+0.00104} = 0.001045019 \tag{4.4}$$

Table 4.5
Model Calculations for CVE-2016-0016

Term	Variable	Weight	Product
cvss_temporal	3	0.020	0.0609
tag_code_execution	1	0.347	0.3472
vend_microsoft	1	0.843	0.8427
ref_misc	1	0.374	0.3737
ref_ms	1	1.435	1.4352
ref_reference_count	2	0.116	0.2072
vuln_count	12	0.114	1.4166
any_exploits	1	4.284	4.2843
in_exploitdb	1	−1.976	−1.9761
Bias	1	−6.630	−6.6305
Log odds			0.3614

CVE-2016-1499 has an estimated probability of exploitation of 0.104%, not very likely.

Applying the same formula to the log odds from CVE-2016-0016 (which is 0.3614), gives us roughly a 58.4% probability.

From the probability we can develop a score. Probabilities operate in the space between 0 and 1. One option for conversion is to simply project into the existing 0–100 scale by multiplying the probability by 100. Another option to explore is the "points to double the odds" method used in the FICO credit score. In that approach, you would scale the probability to a point system in which the probability doubles over a certain span of points.

4.2.3.4 Communicating the Results

There are various methods for communicating the probabilities and scores. Here are a few options.

Incremental probabilities. Using CVE-2016-1499 as an example (Table 4.6), one method is to back into incremental probabilities by adding in one variable at a time.

But care must be taken to not assume that "ref_misc" or any of the other fields have a static and linear effect on the probability. The ordering of the variables will affect the "Change" column above since it's changing in log space. This may be problematic if someone tries to compare across different vulnerabilities and it may raise more ques-

Table 4.6
Incremental Probabilities for CVE-2016-1499

Term	Change	Probability
Bias/intercept		0.132%
cvss_temporal	+0.023%	0.155%
cvss_authenticationRequires single instance	−0.062%	0.093%
integrity_impactNone	−0.027%	0.066%
tag_authenticated	−0.021%	0.045%
ref_confirm	+0.013%	0.059%
ref_misc	+0.026%	0.085%
ref_reference_count	+0.020%	0.105%

tions than they answer ("why is 'ref_misc' 0.02% here and 0.04% there?").

Top influencers. Using CVE-2016-0016 as an example, the major influencers for the rating include:

- A published exploit exists for this CVE;
- Several organizations have reported this vulnerability being present;
- The vendor is reportedly Microsoft.

This solution is relatively scalable (you would just have to write text around each variable), but it lacks any quantification metrics and may be seen as evasive.

Top influencers by proportion. This can be a good compromise. Each weight exists in log space, so if we treat them as independent but as part of the larger proportion, we can estimate its overall influence. This may be a bit confusing though, as it's another percentage, but not directly related to the percentage listed as the probability of exploitation. Let's again look at CVE-2016-0016 (Table 4.7).

So the major influencers would be:

- A published exploit exists for this CVE (+39% influence);
  - In exploit DB (-18% influence).
- Several organizations have reported this vulnerability being present (+13% influence);

Table 4.7
Influencing Variables by Proportion for CVE-2016-0016

Term	Product	Influence	As Percent
cvss_temporal	0.061	0.061	0.56%
tag_code_execution	0.347	0.347	3.17%
vend_microsoft	0.843	0.843	7.70%
ref_misc	0.374	0.374	3.41%
ref_ms	1.435	1.435	13.11%
ref_reference_count	0.207	0.207	1.89%
vuln_count	1.417	1.417	12.94%
any_exploits	4.284	4.284	39.15%
in_exploitdb	−1.976	1.976	18.06%

- The vendor is reportedly Microsoft (+13% influence).

Note the listed percentages are "percentage of influence" and are not summed to the probability of exploitation. They represent the "percentage of influence" over the estimated probability. Another option is attempting to convert these percentages to fractions. Instead of saying "39% influence," you could round to 2/5ths, so "2/5ths of overall shift."

4.3 DESIGNING A NEURAL NETWORK

This section outlines how possible inputs to the model were evaluated, selected, and eventually included in or excluded from the final model. This includes transformations and special considerations for training and testing splits over time.

We'll also detail how we arrived at a neural network model, including an exploration of the variety of architectures that were used. Finally, we'll outline how we transform the model predictions into the final scores. This includes transformations and scaling of the prediction values and incorporation of volume data into the exploits.

4.3.1 Preparing the Data

Overall, 309 possible transformed variables were available for possible modeling. Ideally, we would want to evaluate all combinations of inclusion and exclusion of various variables, especially into a general model like a neural network. Unfortunately, this would require 2,309 different model evaluations (more than the number of atoms in the universe).

For this reason, we break the variables into 14 different classes and evaluate each class individually. Even this more basic number would require an intractable number of model evaluations, but we address how to sample from the space of possible models somewhat uniformly later.

The following list of classes focuses on whether the class of variables was included or not. In cases where the classes may not be "include or don't," we provide a list of possible ways to include variables or not.

Exploitation variables. Whether a CVE has known proofs of concept (POCs) or exploitation code in the wild. Because these generally send a strong signal, we always want to include at least some variant. We break this into two options:

- *All:* Include all possible sources of exploitation as individual binary variables;
- *Any:* Include one binary variable indicating a POC or exploitation code exists anywhere

Vulnerability counts. How prevalent a CVE is in an environment, as a count of occurrence in that environment.

- *Total:* Include the total count of open and closed occurrences;
- *Both:* Include both open and closed counts individually;
- *None:* Do not include any vulnerability counts.

Tags. Whether to include variables derived from the Natural Language Processing tagging system.

Vend. Whether to include vendor information or not.

CVSS. What CVSS information should be included. Note we only evaluate the data used to create the CVSS scores, not the derived scores in the model. If the derived scores correlate with the likelihood of exploitation, a sufficiently complex model should be able to recombine the base values in a similar way.

- *All:* Include all possible CVSS variables, including effects to confidentiality, integrity, and availability (CIA), and various difficulty assessments;
- *CIA:* Only include information on how a CVE affects CIA;
- *None:* Do not include any CVSS information.

CVSS temporal. Whether to include information contained in the CVSS temporal score.

References. Information contained in a CVE's references. Because the number of references was such a strong indicator of exploitation in previous efforts, always include this count. Test whether inclusion of other reference information in addition to count data is effective.

4.3 DESIGNING A NEURAL NETWORK

CWE. Whether CWE information should be included.

CPE. Whether CPE information should be included.

CVE age. Whether we should include the CVE age in the model.

All data should be transformed beforehand. In the case of our example, the count data was first scaled using the log(1+x) transform, and then scaled between 0 and 1. The only exception was CVE age, as this was not heavy tailed and therefore did not require the log transformation first.

Some other considerations before beginning:

Age of the CVEs. Selecting which CVEs to include in the training data is an important consideration. Very old CVEs may behave significantly differently than newer ones, reducing performance on newer data. So we need to decide how far back in time we need to look for training data. In this example, we test using data from 1999 up to 2019.

Year a particular CVE falls in. Because CVEs are sometimes allocated in a year but not published until much later, we can either make these selections based on the publication date or by the year in its CVE ID. We test both possibilities for all the years provided.

Training, validation, and testing. The data span from 1999 up to November 25, 2019. The testing set is static and includes all vulnerabilities published (or IDed) between November 25, 2018 and November 25, 2019. Training data is vulnerabilities published after January 1 of the selected data start date (from the prior section) up until November 25, 2018. In that time period we reserve 10% of the data for validation. This data is used for tuning the training process, in particular early stopping.

Volatility. A time-based model requires careful selection of training, validation, and testing data. In particular, a random split between these different data sets generally would mean that the inclusion of additional variables would almost certainly increase model performance. However, variables that may contribute to exploitation in the past may not in the future.

To evaluate whether a variable will consistently contribute to model performance, we take a simple approach. For each variable, we create a rolling one-year window across the training data. On each window, we create a simple logistic regression for the variable against exploitation. We measure the direction and statistical significance of the coefficients for the regression over time. We then measure the number of times the sign of the coefficient changes direction over the course of the rolling window. Many directional changes indicates a volatile variable that may not necessarily be relied upon for future predictions.

We compute the quantile function on the number of flips across all variables. We use that quantile function as a way to include or include variables at specific cutoffs, and test six different cutoffs:

- Include all variables;
- Exclude only the top 10% most volatile variables;
- Exclude the top 25% most volatile variables;
- Exclude the top 50% most volatile variables;
- Exclude the top 75% most volatile variables;
- Only include variables that have no coefficient sign changes.

4.3.2 Developing a Neural Network Model

Generally, we are focused on the area under the precision recall curve (AUC PR or just PR) for evaluating neural network models. This provides a good metric that indicates the model's performance without being skewed by the highly imbalanced data set.

Before settling on a neural network model, we explored several other possibilities, including regularized linear models (similar to EPSS), random forests, and decision trees. None of these alternatives had performance similar to a very simple, untuned neural network model. We therefore focused only on a variety of neural network models.

4.3.2.1 Neural Network Architecture

Before training begins, we need to make a number of decisions about the architecture of the network. Every decision impacts performance. We used a simple dense neural network model with several hidden layers. Because the data is nonsequential and nonspatial, we did not investigate other architecture types like recurrent and convolutional networks, respectively.

Future work might focus on creating residual connections or using auxiliary outputs on deeper networks, but given the performance we observed, it is unlikely to create significant performance increases. Here are the decisions we made and why.

Layers. We investigated a number of different network depths from one to three. Zero hidden layers would indicate a simple linear model, which we had already established as performing significantly worse than a network with hidden layers. Early experiments on "deeper" networks (not "deep" networks, a moniker that should be reserved for networks with more than a dozen hidden layers, although there is some debate about what the cutoff should be) indicate that four- and five-layer networks did not perform particularly better than shallow networks and so weren't included in the final evaluation.

Size. The sister parameter to the number of layers is the number of hidden units contained within those layers. We examined powers of two from 64 to 1024. Powers of two fit better into graphics processing (GPU) memory and confer significant performance advantages.

Wide, narrow, or constant. It is possible that each subsequent hidden layer in the network could be of different sizes. We tested this via either doubling or halving the number of hidden units in each subsequent layer, or alternatively keeping them constant.

Activations. We explored several types of activation functions in early experiments but found little performance difference so settled on the popular *rectified linear unit* (ReLU) activation.

Regularization. Often to prevent overfitting, networks require a type of regularization to prevent them from settling into a deep local minimum. These can include "dropout" in which random inputs and internal signals are ignored during training. This also includes modifying the loss function to include the magnitude of the weights as a penalty. In early experiments, none of these methods significantly mitigated overfitting as well as simply stopping when the validation loss failed to improve.

Smoothing. Another form of regularization that is particularly effective at ensuring that the network is not overfit into a single minimum is label smoothing. This process works by converting 0s into a small positive value epsilon and 1s into 1-epsilon. We tested a number of different epsilon values including 0, 0.0001, 0.001, 0.01, and 0.1.

Stopping. During training, constantly declining loss on the training data can occur by pushing the model deeper into the extreme of a local minimum. To combat this, we held out 10% of validation data that's stopped once the validation data failed to improve on each subsequent pass of the data.

4.3.3 Hyperparameter Exploration and Evaluation

Given all the possibilities for data inclusion and model architectures, there are 1,128,038,400 unique possible models. Given that each model takes approximately 1–10 minutes to fit, depending on these hyperparameters it would take approximately 120 years of computation time. For this reason, we need to sample from the space of models.

We do this via Latin hypercube sampling. This technique allows us to ensure that we explore all possible values in each individual parameter, and that we explore as many low-level combinations of parameters (two, three, etc.) as possible. We sampled 5,000 models this way and evaluated each variable based on its AUC PR.

After fitting the 5,000 models, we examined a simple bivariate beta regression model comparing the AUC PR of the model versus the various hyper parameter values. We use this simple model to select the best hyperparameter. When there was no significant impact on the model performance, we selected the more parsimonious of the models (i.e., not including variables or selecting the smaller architecture).

4.3.3.1 CPE

We saw slightly better performance when CPE values were not included (Figure 4.7).

4.3.3.2 CVE Age

Including CVE age did not significantly affect model performance (Figure 4.8).

Whether selecting variables based on CVE ID or CVE publication date did not impact performance (Figure 4.9).

4.3.3.3 CVE Start Year

The cutoff used for the beginning of the data significantly impacts performance, but unfortunately not in a completely clear way

4.3 DESIGNING A NEURAL NETWORK

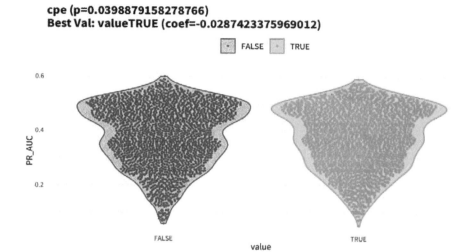

Figure 4.7 Performance when CPE values were not included. (© 2018 Kenna Security. Reprinted with permission [6].)

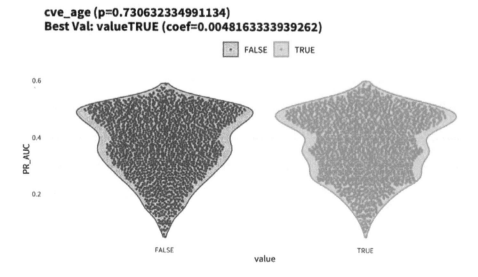

Figure 4.8 Performance when including CVE age. (© 2018 Kenna Security. Reprinted with permission [6].)

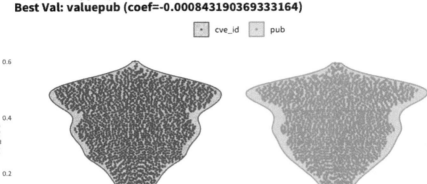

Figure 4.9 Performance when selecting variables based on CVE ID or CVE publication date. (© 2018 Kenna Security. Reprinted with permission [6].)

(Figure 4.10). The best performance was achieved using a cutoff of 2009 and 2012.

However, there was little difference in the overall distribution of models. We selected 2009.

4.3.3.4 CVSS

Interestingly, including CVSS variables negatively affected performance, so we chose to exclude them from the model (Figure 4.11).

4.3.3.5 CVSS Temporal

Including the CVSS temporal score had no effect on model performance (Figure 4.12).

4.3.3.6 CWE

Including CWE variables did not impact model performance (Figure 4.13).

4.3.3.7 Exploit Variables

As expected, including variables with information on exploitation positively impacts performance. We include all exploit variables independently (Figure 4.14).

4.3 DESIGNING A NEURAL NETWORK

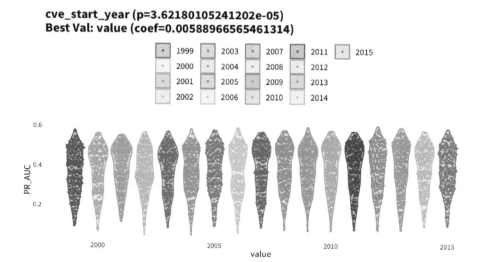

Figure 4.10 Performance with different cutoff years for the beginning of the data. (© 2018 Kenna Security. Reprinted with permission [6].)

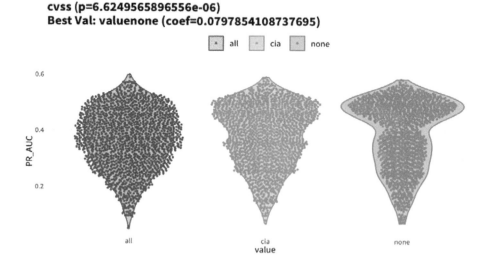

Figure 4.11 Performance when including CVSS variables. (© 2018 Kenna Security. Reprinted with permission [6].)

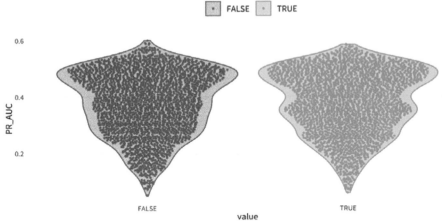

Figure 4.12 Performance when including the CVSS temporal score. (© 2018 Kenna Security. Reprinted with permission [6].)

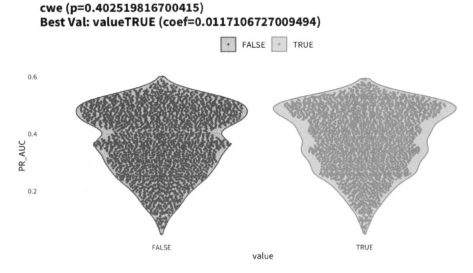

Figure 4.13 Performance when including CWE variables. (© 2018 Kenna Security. Reprinted with permission [6].)

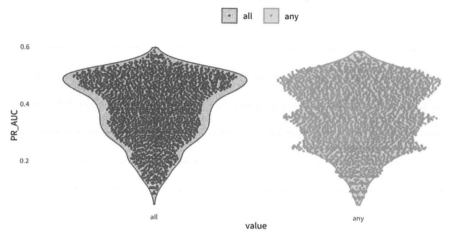

Figure 4.14 Performance when including variables with information on exploitation. (© 2018 Kenna Security. Reprinted with permission [6].)

4.3.3.8 Product Variables
Including product variables improves performance (Figure 4.15).

4.3.3.9 Reference Variables
Including reference variables improves model performance (Figure 4.16).

4.3.3.10 Tags
Including tags variables does not impact performance (Figure 4.17).

4.3.3.11 Vendor
Including vendor information improves performance (Figure 4.18).

4.3.3.12 Vulnerability Count
Including count information positively impacts performance. Including both counts results in the best performance (Figure 4.19).

4.3.3.13 Volatility Cutoff
Dropping the 25% most volatile variables produces the best performance (Figure 4.20). Note here that 0.0 indicates dropping no variables, and 1.1 indicates including all variables.

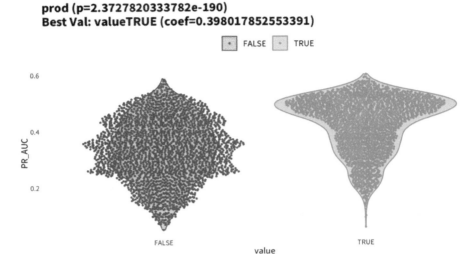

Figure 4.15 Performance when including product variables. (© 2018 Kenna Security. Reprinted with permission [6].)

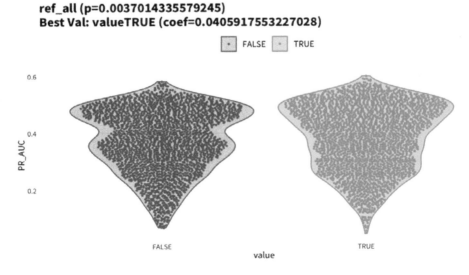

Figure 4.16 Performance when including reference variables. (© 2018 Kenna Security. Reprinted with permission [6].)

4.3 DESIGNING A NEURAL NETWORK

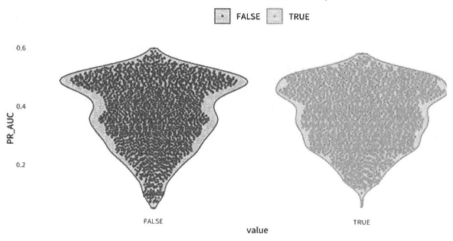

Figure 4.17 Performance when including tags variables. (© 2018 Kenna Security. Reprinted with permission [6].)

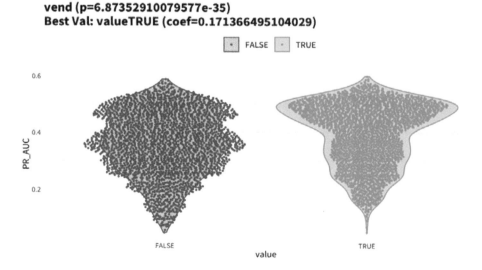

Figure 4.18 Performance when including vendor information. (© 2018 Kenna Security. Reprinted with permission [6].)

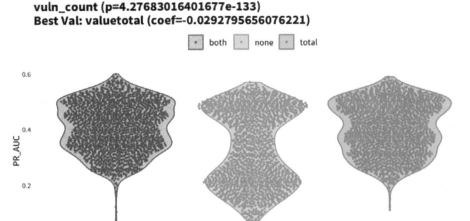

Figure 4.19 Performance when including count information. (© 2018 Kenna Security. Reprinted with permission [6].)

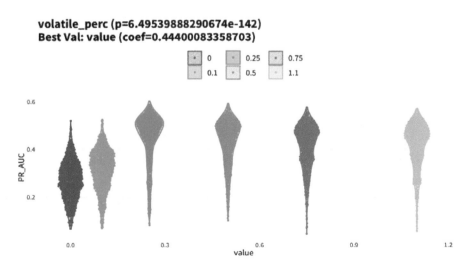

Figure 4.20 Performance when dropping the 25% most volatile variables. (© 2018 Kenna Security. Reprinted with permission [6].)

4.3.3.14 Number of Hidden Layers

The best models had two hidden layers, although three layers was not particularly worse (Figure 4.21).

4.3.3.15 Hidden Layer Size

The size of the hidden layers seemed to make little difference (Figure 4.22). However, the top-performing models always had a size of 512, so although on average there was little difference, in the top models a larger capacity seemed to be needed. Note the values here are logged.

4.3.3.16 Hidden Factor

We found that narrowing or increasing the number of hidden units had only a small effect on performance (Figure 4.23). However, the best values were achieved with no change (hidden factor 1).

4.3.3.17 Label Smoothing

On overall performance, no label smoothing resulted in the strictly best performance. However, we also found that models with no label

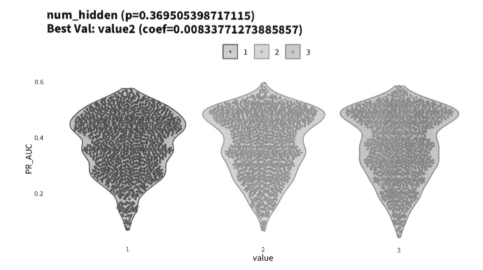

Figure 4.21 Performance with different numbers of hidden layers. (© 2018 Kenna Security. Reprinted with permission [6].)

94 HOW TO BUILD A DECISION ENGINE TO FORECAST RISK

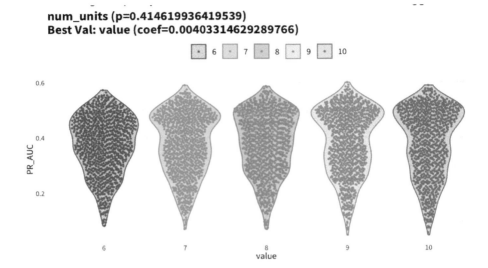

Figure 4.22 Performance with hidden layers of different size. (© 2018 Kenna Security. Reprinted with permission [6].)

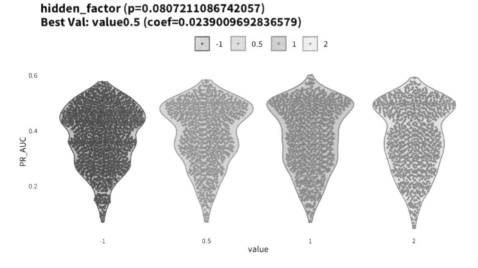

Figure 4.23 Performance when narrowing or increasing the number of hidden units. (© 2018 Kenna Security. Reprinted with permission [6].)

smoothing tended to produce predictions that were very bimodal (either almost 1 or numerically 0), while a small amount of label smoothing (epsilon = 0.01) provides a smoother distribution of values.

4.3.4 Scoring

While most of the above decisions are data-driven, some of the details of the final model are somewhat more flexible depending on the desired properties of the final score. Where possible, we indicate which portions of the score to which we can make changes.

4.3.4.1 Score Scaling

The predictions provided by the model are skewed toward small values, given the imbalance in the data. The distribution of predictions for exploited and non-exploited variables can be seen in Figure 4.24.

To combine it with volume information to create the final score, we need a less skewed distribution. We can achieve this through a log transformation and a scaling between 0 and 1 (Figure 4.25).

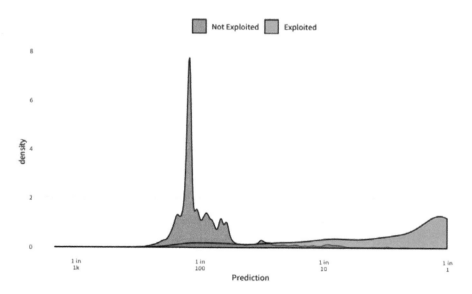

Figure 4.24 Distribution of predictions for exploited and nonexploited variables. (© 2018 Kenna Security. Reprinted with permission [6].)

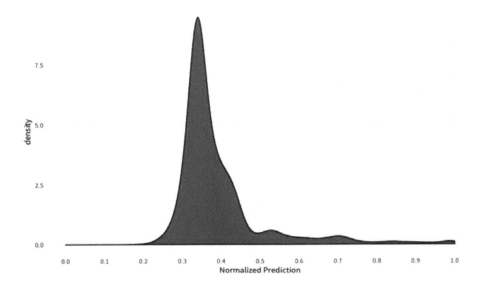

Figure 4.25 Normalized prediction achieved through a log transformation and a scaling between 0 and 1. (© 2018 Kenna Security. Reprinted with permission [6].)

This does concentrate the normalized tightly around 0.325. Other transformations might be used to spread the distribution more evenly throughout the space or give it a different center.

4.3.4.2 Volume Scaling

Incorporating exploited vulnerabilities via their volume data is an important step. We found that exploit volume is extremely heavy tailed. To arrive at something resembling a reasonably smooth distribution, two log transformations are needed (Figure 4.26).

It is possible that we may want to provide different transformations for different types of CVEs.

4.3.4.3 Combining Scores

We combine prediction and volume via the Euclidean norm. Because the volume and exploitation prediction are normalized between 0 and 1, this value is bound between 0 and the $\sqrt{2}$. If we want to bound the scores between 0 and 1 (and later 0 and 100), we can take a few approaches.

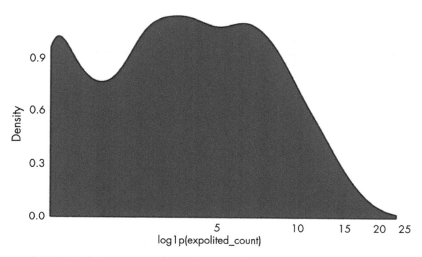

Figure 4.26 Exploit volume with two log transformations. (© 2018 Kenna Security. Reprinted with permission [6].)

None. It is unlikely a vulnerability would max out likelihood of exploitation and a volume beyond what we have seen before seems somewhat unlikely. We could do no normalization on this final value.

Normalize by the theoretical value. That is, divide the above by the $\sqrt{2}$. This would ensure that even if we had a maximum volume and prediction vulnerability in the future, it would have a score of 1, and not more.

Empirical normalization. That is, scale the final score between the min and max value. Because this value is less than $\sqrt{2}$ (1.3533), this may increase the likelihood in the future of a value greater than 1.

Finally, we could avoid any range issues by using any of the normalizations above and passing the final score through logistic function as described in the engineering spec. However, this caps the lowest and highest values between approximately 5 and 90.

The different normalizations can be seen in Figures 4.27 and 4.28.

4.3.4.4 Comparison to Existing Scoring Model

To show the value in this new risk-based vulnerability management program, we can plot the efficiency and coverage of your current scoring model and the predicted probability of this model. We'll cover

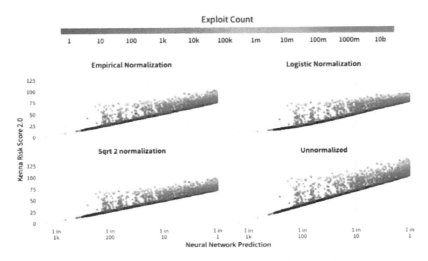

Figure 4.27 Different normalizations applied to the volume and exploitation predictions. (© 2018 Kenna Security. Reprinted with permission [6].)

efficiency and coverage in greater depth in Chapter 5 as a way of measuring performance.

Sometimes comparing two models can be difficult—one might have manual adjustments or another difference that prevents an apples-to-apples comparison.

For example, when Kenna Security measured this model against an older version of its Kenna score, the two didn't exactly line up. The existing Kenna score had a manual adjustment if the CVE was actively being exploited in the wild, although the model applies CVEs found in Proofpoint that were not part of that manual adjustment. Even with the manual adjustment on the Kenna score, let's see how they compare in Figure 4.29.

We can see the initial bump that the Kenna score gets from the manual adjustment. The Kenna score is highly precise right out of the gate. The model output also has a bump out of the gate, but that's just a fluke from a very small number of CVEs getting high probabilities.

Rather than just a straight coverage/efficiency plot, a better way to compare these is to include a comparison at the same level of effort. In other words, out of the first 500 CVEs recommended to remediate, how does the model output compare to the Kenna score? How about at 1,000 or 5,000 recommended CVEs? (See Figure 4.30.)

4.3 DESIGNING A NEURAL NETWORK

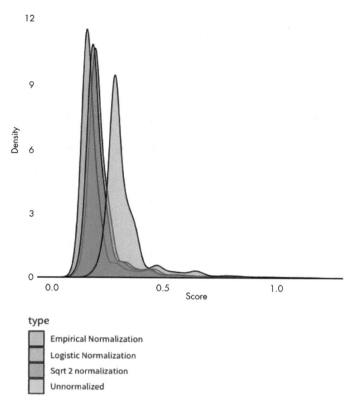

Figure 4.28 Different normalizations applied to the volume and exploitation predictions. (© 2018 Kenna Security. Reprinted with permission [6].)

The dotted line at 0% represents no change over the existing Kenna score, when the line is above it represents an improvement and below represents a decrease. It's easier to see the benefit of the model output over the level of effort in this type of visualization. We get peak efficiency improvement around 420 at 22% improvement, and coverage peaks around 7,000 CVEs with a 13% improvement in coverage. On average, the model output is a 6.2% increase in coverage and a 3.3% increase in efficiency.

The important part here is that efficiency and coverage are metrics that allow us to compare two models and reasonably forecast the impact of switching to one. Machine learning algorithms themselves have no value—whether it's a neural net or a coin flip tied to a lava

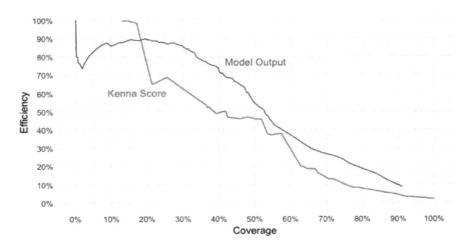

Figure 4.29 Comparing scoring models by coverage and efficiency. (© 2018 Kenna Security. Reprinted with permission [6].)

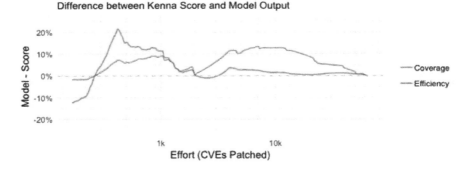

Figure 4.30 Comparing scoring models while factoring in effort. (© 2018 Kenna Security. Reprinted with permission [6].)

lamp deciding which vulnerabilities to fix doesn't matter—what matters is that the outcomes of the model are consistently increasing the efficiency of the vulnerability management program.

4.3.5 Future Work

While this model represents the best we have available to us today based on the available data and current technology, there are of course additional features and functionality that would benefit security and operations teams in assessing risk and addressing the most critical vulnerabilities.

The model shown in Figure 4.30, for example, currently ignores the element of time. There is more work to do to track actively exploited vulnerabilities as opposed to those that have been exploited in the wild at some point in time. For example, CVE-2018-11776, the Apache Struts vulnerability behind the Equifax data breach, moved from a CVE to a proof of concept to a weaponized proof of concept in a matter of 48 hours. We need more event sources for fast-moving vulnerabilities like Struts when we don't yet have an event but know it's happening now.

Along the same lines, vulnerabilities with the potential to move fast are higher risk. We saw this with WannaCry, which was risky before EternalBlue emerged. We also need to factor in the attack path. Spectre and Meltdown, for example, illustrate that for some vulnerabilities, the risk is extreme in some cases, but not others.

The model we mapped out also does not have any differentiation between "I don't know" and "No." There could be some more research to create simpler models to step toward this more robust model. For example, explore a model without CVSS or other NVD variables that would enable you to generate a rating prior to NVD data on a new CVE. Or maybe look at models for non-CVE (appsec) vulnerabilities.

We may be able to reduce the feature list to only those that show a split across vulnerabilities exploited in the wild and not exploited. This could replace the cutoff where we removed variables in the bottom half of CVE counts.

Now that we have a decision engine, let's look next at how to evaluate its performance.

References

[1] MacKenzie, D., *An Engine, Not a Camera*, Cambridge, MA: MIT Press, August 29, 2008, https://mitpress.mit.edu/books/engine-not-camera.

[2] *National Vulnerability Database*, CVE-2020-0787, March 12, 2020, https://nvd.nist.gov/vuln/detail/CVE-2020-0787#match-5328563.

[3] Synopsys, *Automate AppSec Triage with Machine Learning*, 2021, https://www.synopsys.com/software-integrity/resources/ebooks/use-machine-learning-for-appsec-triage.html.

[4] Kenna Security, *Logistic Regression Model Documentation*, 2018.

[5] https://drive.google.com/open?id=1Exx0V9U818Y_meWc2phlqe8Ta5Hm-5bQs.
[6] Kenna Security, *Neural Network Model Documentation*, 2020.

5

MEASURING PERFORMANCE

When Henry Ford switched on the first moving assembly line in 1913, he introduced efficiencies and automation that not only changed manufacturing forever, but lowered the price of the Model T enough that it would quickly become ubiquitous.

Ford pushed car manufacturing into motion by borrowing from meat-packing plants, canneries, flour mills, and other industries that used conveyor belts. Assembling an entire car in one spot was time consuming and inefficient, but, he hypothesized, he could speed production by bringing the cars to the workers.

In 1912, the year before the assembly line was launched, Ford's traditional manufacturing process produced 68,773 Model Ts. In 1913, the moving assembly line produced 170,211. Four years later, 735,020 Model Ts rolled out of the factory, more than nine times the number from just five years earlier [1].

That acceleration is the result of incremental improvements over time. The assembly line meant workers performed discrete tasks that not only saved unnecessary movements, but the specialization also increased quality and reduced errors and flaws [2]. Ford began to use parts that were automatically stamped together by machines, saving even more time and increasing quality.

Ford was able to continue refining the process, saving time, increasing capacity, and producing more cars, because he was able to build up enough data to measure the factory's performance.

One of those metrics was assembly time per car. Before the assembly line, a Model T took 12.5 hours to assemble. After the assembly line it took just 93 minutes. Just like that, he went from producing 100 cars per day to 1,000 [3]. But until he had assembled enough cars and measured his workers' assembly rates, he couldn't say what an acceptable assembly time per car might be.

In any field, measuring performance is key to consistent improvement. Over time, a stable model of data develops that allows you to identify metrics that define success based on data. By focusing on the right metrics, you can start to define which processes can be automated, and then find new efficiencies.

Just as Ford needed enough data to define expectations and improve performance, in vulnerability management we need to accumulate and analyze performance metrics to ensure we're hitting our goals and mitigate risk within an acceptable time frame.

In this chapter we'll delve into how to define a worthwhile metric, why performance matters, and which metrics you should be tracking to gauge your performance in lowering your organization's risk.

5.1 RISK VS PERFORMANCE

A large insurance customer was having great success in adopting risk-based vulnerability management. They went so far as to incentivize IT operations to drive risk down by bonusing them on staying below a certain level of risk. At first, this seemed brilliant—and it was; the results were there. But over time, something curious started to occur—risk would stay around the level they strived to get to, but every once in a while it would spike up above the chosen tolerance. Sometimes, this spike would happen right around the time bonuses were measured (end of the quarter), and the IT operations folks took note. Since risk itself is dependent on factors such as time, attackers, the infrastructure, and the vendors, a lot was outside of their control. Before we look at what makes a metric good, we need to understand the relationship between performance and risk.

Risk combines the likelihood of an event with the cost of that event occurring. Often that's not a single number, but a range of values. Measuring risk is about trying to quantify what threats your organization faces and the likelihood you'll experience their consequences.

Much of the last chapter covered ways to think about risk and models that allow us to compensate for randomness and unpredictability.

Performance, on the other hand, doesn't measure risk, but your level of success in mitigating and managing risk. It measures the actions you've taken to achieve your goal of mitigating the risks you face. Performance gives you benchmarks for improvement and better visibility into your exposure to risk.

How you perform on various measures affects your risk either directly or indirectly. While the two are distinct concepts, in the field of vulnerability management, they're intertwined. Better performance should equate to lower risk.

But that's only possible if you're using the right metrics.

5.2 WHAT MAKES A METRIC "GOOD"?

Metrics support decisions. Good metrics lead to good decisions. They also offer an objective way to assess the merits of different remediation strategies.

But what exactly makes a metric good? Which metrics should you use to gauge your success at fixing vulnerabilities and lowering risk?

Deciding which vulnerabilities to remediate is a daunting task. In a perfect world, you could remediate every vulnerability as soon as it's discovered. In the real world, with thousands of new vulnerabilities every year multiplied across disparate, ever-changing assets, you need to prioritize which vulnerabilities to focus on first.

You also need a way to determine that you're making the right decisions over time. There are many different remediation strategies to choose from as well as many different measurement techniques to evaluate remediation strategies.

In this section, we're going to look at the criteria behind every good metric and introduce you to several measures of performance. Some you've likely heard of before and are quite common but ultimately don't hold up as effective metrics. Hopefully we'll also introduce you to some new metrics and models that can help you better understand the trade-offs inherent in fixing vulnerabilities and lowering risk, as well as how to strike the best balance for your particular organization.

5.2.1 Seven Characteristics of Good Metrics

5.2.1.1 Bounded

A bounded metric has finite limits, whether naturally or imposed on it. Examples of bounded metrics include percentages of a whole and true-false scenarios, both of which have a set scale.

Unbounded metrics aren't useful in the decision-making process. For example, if you're measuring performance by the metric "days to patch," you might aim to lower the number of days it takes to apply a patch by increasing the number of employees working on the problem. But basing hiring decisions on an unbounded metric is not recommended. If it takes 100 days to patch with three employees, will hiring three more lower it to 50? If a vulnerability remains unpatched and the "days to patch" is infinity, how many employees do you need to hire? Without clear bounds, the metric quickly falls apart.

Bounded metrics exist on a scale that allows you to define the state of no people hired or many people hired or the state of I'm patching well or I'm patching poorly.

5.2.1.2 Scales Metrically

The difference between two values of any metric should offer meaningful information.

For example, percentages scale metrically because the difference between any two values is consistent and you can easily determine whether and how much the values have changed. If you've remediated 10% of vulnerabilities in your organization, and then improve to 15% of vulnerabilities, that's a measurable difference that's consistent and obvious.

If, however, you're using a scoring system and can't articulate how much better or worse a particular score is, that's not useful.

5.2.1.3 Objective

Objective metrics are resistant to changes in input data. They should be easily tested and reproduced.

One simple example is time. The date and time when you discovered a vulnerability or patched a vulnerability isn't going to change and everyone can agree on and verify those numbers.

5.2.1.4 Valid

A metric is valid if it successfully measures what it sets out to measure [4]. That is, it has a measurable output that consistently represents the difference or change between two values.

For example, in a metric that tries to measure exploitability, the metric should rise if there are more exploits. If it doesn't, it's not a valid metric.

5.2.1.5 Reliable

Reliable metrics generate the same outputs from the same inputs.

5.2.1.6 Context-Specific; No Gaming

To explain context-specific metrics, we need to quickly introduce the concept of Type I and Type II metrics. Type I is a metric in a controlled environment. It ignores the threat environment to generate a base rate to which you can compare a threat environment metric. CVSS, for example, is descriptive and offers rich data, but it says nothing about how the vulnerability interacts with the threat environment.

Type II metrics, on the other hand, take into account the larger context around what hackers are doing. For instance, does a particular vulnerability have a successful exploitation? What is the percentage of infected machines in your environment?

Ultimately you need both Type I and Type II metrics. Type I should not influence decisions or policy, but they're useful for generating Type II metrics with input or information from that threat environment. That's how to use threat intelligence: Take a number of Type I metrics that exclude the real-life threat environment, and control for the occurrence rate, and that tells you about the behavior of your users or control systems. Overlay that with threat intelligence and you're generating Type II metrics about the threat environment.

A context-specific metric takes into account the larger threat landscape and attacker behavior, both to reflect the reality in which remediation is taking place and to prevent gaming the metrics so the numbers look good even as your organization's risk increases.

If your adversary changes their behavior, that voids the validity of a metric that's not context-specific. A good metric will respond to changes in attacker behavior, either by factoring in events and actions, like whether an exploit for a vulnerability exists, or by augmenting data with threat intelligence.

The more context you have, the more data you can provide, and therefore the more difficult it is to game the metric.

5.2.1.7 Computed Automatically

Unless you have an army of foot soldiers, a manually computed metric won't be helpful in measuring performance. Manual analysis is possible, but a metric that relies solely on manual computation is useless against a real-time attacker. As much as possible, focus on the many worthwhile metrics that can be computed automatically.

5.2.2 Evaluating Metrics Using the Seven Criteria

Let's look at a few examples of common vulnerability remediation metrics in the context of these seven criteria to see how they hold up.

5.2.2.1 Mean Time to Incident Discovery

This metric aims to assess the performance of the incident response (IR) team by calculating how quickly, on average, they discover an incident. But it's a terrible metric.

It does scale metrically—20 days to discovery is twice as bad as 10 days. It's objective in that the time and date when the incident occurred and when you discovered it are stable numbers that aren't arguable. It's reliable because if there are two numbers, you'll get the same metric every time, and it is context-specific because it tries to describe incident response.

But ultimately this metric won't work, for several reasons. First, it's not bounded. You might never discover an incident, making the time frame infinity and skewing the results. You could compensate for this by limiting the time frame to a target of 100 days, for example.

But mean time to incident discovery is also not valid. The number of days to discovery is not something that can necessarily be counted because it could be obfuscated by a number of actions on the part of the attacker that would make the timeline unclear. In addition, as the metric increases, it doesn't necessarily describe the performance of the IR team because there are any number of other factors that could influence the number of days. Teams in different enterprises could be executing the same actions faster or slower based on what's happening in their control environment, what processes are at stake, and what logs they have access to.

But worst of all, mean time to incident discovery can't be computed automatically. If you could compute it automatically, you would have zero for your incident response time. You would have already responded to the incident. You would have to manually calculate it after the fact, discovering when the incident happened based on the new rule sets generated from the incident. And if that's a part of the input into the metric, you can't really measure adherence to the metric over time because different people could adhere to it differently.

Three strikes and mean time to discovery is out.

5.2.2.2 Vulnerability Scanning Coverage

This metric aims to track the percentage of assets you've scanned for vulnerabilities. Let's see how it performs.

It's bounded because it's a percentage, so differences can easily be measured. For the same reason, it scales metrically. You know that scanning 75% of assets is 50% better than scanning half of them.

It can be automatically computed because once you launch the scanner and have your asset inventory, it's possible to scan those assets automatically. It's context-specific in that it's trying to describe a particular process that nothing can obfuscate.

We run into some complication, however, when we look at whether it's objective, valid, reliable, and can be gamed. It would be nice if we could say that scanning is objective since it's a binary choice about every asset—it has either been scanned or it hasn't. But scans themselves can introduce numerous issues, such as whether the scan is authenticated or unauthenticated. As a result, it may not be valid and reliable.

Often one metric will be insufficient, and five may overload decision-making, but the correct combination can offer both context and feedback. In this case, an additional metric to watch could be the percentage of scans authenticated, which means it's valid. Similarly, we can measure false positives by scanning the same assets with multiple technologies, introducing a metric like false positives/false negatives of auth versus unauth scanning. By sampling, say, 100 assets, you could create an internal measure of how a particular scan will affect the data.

Vulnerability scanning coverage is clearly a good metric.

5.2.2.3 CVSS for Remediation

Remediating vulnerabilities over a certain CVSS score is a common strategy. But does CVSS for remediation offer a good metric for measuring performance?

It's certainly bounded, with the range of scores from 0 to 10. And it's reliable—the same inputs generate the same outputs. It can be computed automatically. But otherwise it's a mess.

First, it doesn't scale metrically. Is a vulnerability with a CVSS score of 10 twice as bad as one that scores a 5? How much more critical is a vulnerability that's rated 9.7 versus one that's 9.3? We don't know, which makes it difficult to act on this information.

While CVSS itself is objective, mapping those inputs to a score is subjective. CVSS for remediation is not valid in that it doesn't describe the order in which to remediate the vulnerabilities; it simply describes each vulnerability. It's not context-specific because it doesn't take into account the threat environment, and you could easily game CVSS by saying you don't have information about a certain input that would affect the scores—this happens quite frequently.

This one is more complicated, and a mixed bag, but ultimately CVSS for remediation is not a good metric.

5.2.3 More Considerations for Good Metrics

These seven criteria offer a standard set of rules to assess the metrics you're using to measure performance. But note that none of these criteria is perfectly achievable. Whether a metric is truly objective can be arguable. Automation exists on a spectrum.

A metric doesn't have to be perfect, so don't let the perfect be the enemy of the good. Use these criteria as a goal to aim for, but consider allowable exceptions. For example, if you have a metric that's not computed automatically but is computed easily, that might be fine for your organization. In terms of reliability, you should get roughly the same measure every time over time, but it also might vary. Some measures are going to be noisier than others but might still provide some signal.

Just as important as the seven criteria is that each metric you use makes sense for your operation. Since there is no such thing as a perfect metric, aim instead for a useful one.

We talked earlier about avoiding gaming. It's important to keep in mind Goodhart's law: "Any observed statistical regularity will tend to collapse once pressure is placed upon it for control purposes," which was broadened by anthropologist Marilyn Strathern to the summary you may be more familiar with: "When a measure becomes a target, it ceases to be a good measure."

When you decide to optimize for a particular metric, people are going to find ways to optimize without actually achieving the intended outcome. This goes back to the relationship between risk and performance. Even if you have a metric that you think is tightly coupled with risk, you might improve that measure without actually reducing risk.

In the perfect world, we would avoid gaming completely, but in the real world, you have to keep an eye on it and do your best to align incentives with the outcomes you're trying to achieve, which in the case of vulnerability management, should be to lower your overall risk.

One way to avoid gaming is to eliminate incentives, economic or otherwise, based on a metric. Metrics should, in some ways, be passive. We should be aware of metrics and they should improve when we take the right actions. But the goal should never be to drive them up. The goal should be lowering your organization's risk. The metrics are just a reflection of how well you're doing.

5.3 REMEDIATION METRICS

There are a wide variety of metrics you can use to assess how quickly you eradicate vulnerabilities and reduce risk. Let's look at some metrics and models, their limitations and trade-offs, and why the size and capabilities of your organization might not matter when it comes to speed.

5.3.1 Mean-Time-Tos

Some of the most common and widely used metrics are the mean-time-tos. We briefly defined these metrics in Chapter 1. To recap and expand, mean-time-tos attempt to benchmark the time frames for vulnerability remediation by finding the average time it takes to complete various actions.

MTTR: On average, how many days pass between when a vulnerability was discovered and when it was closed?

MTTD: On average, how many days passed between when a vulnerability was present in your environment and when you discovered it?

Mean time to exploitation (MTTE): On average, how many days pass between the discovery of a vulnerability and when it's exploited?

All of these can be useful metrics to track through survival analysis and can give you a way to quantitatively benchmark remediation timelines. But it's important to understand what mean-time-tos tell you and what they don't.

For example, many organizations look at MTTR to gauge their overall remediation performance. But that's a mistake. MTTR is limited in the sense that it's only useful to measure what you've actually fixed. Since organizations fix anywhere from 5% to 25% of their vulnerabilities, 75% of vulnerabilities or more remain open.

But MTTR does allow you to assess, within the vulnerabilities you've remediated, how many days on average they took to fix. Let's say you have 100 vulnerabilities and fix 10 of them. You fixed five in 30 days each and another five in 100 days each, for an MTTR of 65 days. You're missing the bigger picture—that 90% of the vulnerabilities are still out there—but you do have a gauge for those you've remediated.

MTTR and MTTE can help you set SLA time frames that align with your organization's risk tolerance. If your tolerance is medium or high, use MTTR. If your tolerance is low, MTTE may be a better metric.

As we've seen, mean-time-tos can be useful in the right context. As with any metric, it's important to understand what it actually tells you and use it accordingly. And as we'll learn later, the power-law distribution of vulnerabilities may make the averages less important in reducing risk.

5.3.2 Remediation Volume and Velocity

While MTTR offers only a limited view of the actual speed of your remediation efforts, survival analysis offers us metrics that paint a more complete picture of your efforts by including the vast majority

5.3 REMEDIATION METRICS

of vulnerabilities that might never be fixed—and don't necessarily need to be.

Survival analysis is a set of methods to understand the time duration of an event. In our sphere, the event of interest is the remediation of a vulnerability, and it's a useful way to view timelines in vulnerability management.

When you do, you can uncover the velocity of your remediation efforts. Every organization has a different velocity—the speed and direction of its remediation efforts. Velocity gives you a more comprehensive picture of your speed of remediation, taking into account all the vulnerabilities in your environment. Survival analysis also indicates the volume of vulnerabilities that have not been fixed.

Let's again assume an organization observes 100 open vulnerabilities today (day zero) and manages to fix 10 of them on the same day, leaving 90 to live another day. The survival rate on day zero would be 90% with a 10% remediation rate. As time passes and the organization continues to fix vulnerabilities, that proportion will continue to change.

Tracking this change over time produces a curve shown in Figure 5.1 where we see that, on average, it takes firms about a month to remediate 25% of vulnerabilities in their environment. Another two months get them over the halfway mark, pegging the median lifespan of a vulnerability at 100 days. Beyond that, there's clearly a long-tail challenge for remediation programs that results in 25% of vulnerabilities remaining open after one year. The stairsteps you see in the line between those points reflect the fits and starts of pushing patches.

Figure 5.2 offers a simplified view of the data in Figure 5.1 that focuses on the first, second, and third quartiles. It typically takes 26 days for firms to remediate 25% of vulnerabilities, 100 for 50%, and 392 days to reach 75%.

It should be intuitive that wide variation exists around the aggregate remediation timeline from Figure 5.1. We include Figure 5.3— which is admittedly an indecipherable mess—to illustrate how wide that variation actually is among organizations from a sample of 300.

It's plain that they cover the entire spectrum. We chose to color the lines by industry to further make the point. There is no obvious pattern to indicate, for instance, that all firms in a sector cluster to-

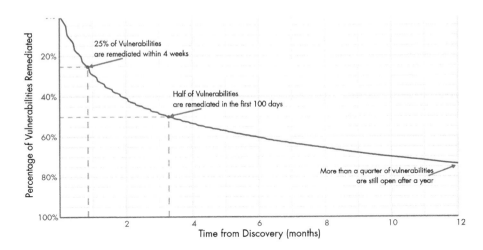

Figure 5.1 Vulnerability survival analysis curve. (© 2019 Kenna Security/Cyentia Institute. Reprinted with permission [5].)

Alternate view of overall vulnerability survival analysis across firms

Figure 5.2 Simplified view of vulnerability survival analysis. (© 2019 Kenna Security/Cyentia Institute. Reprinted with permission [5].)

gether. That's not to say no differences exist among industries at the aggregate level; they do and we'll cover that a bit later.

Why do such extreme variations exist among firms in their remediation timelines? There are a vast range of factors inside and outside these organizations that might impact these results, which in some ways makes these differences not so surprising at all.

We'll dig deeper into the metric of velocity later in this chapter.

5.3.3 R Values and Average Remediation Rates

Measuring remediation velocity brings up two related metrics that are key to measuring performance: asset complexity and remediation capacity.

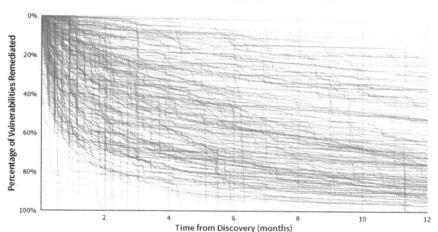

Figure 5.3 Survival analysis curves for 300 organizations. (© 2019 Kenna Security/Cyentia Institute. Reprinted with permission [5].)

Asset complexity refers to the scope, diversity, and intricacy of an IT environment. Remediation capacity measures the proportion of open vulnerabilities a firm can close within a given time frame.

The relationship between asset complexity and remediation capacity is interesting in that complexity and capacity rise in tandem. In other words, a typical organization, regardless of asset complexity, will have the capacity to remediate about one out of every 10 vulnerabilities in their environment within a given month. That seems to hold true for firms large, small, and anywhere in between.

We built a regression model to show the relationship based on data from 300 organizations of varying size. Since we don't have access to all of the information about each organization's asset complexity, we used the total monthly average number of vulnerabilities as a proxy for complexity. Our logic (which is backed by the data) is that more complex environments will have more assets with more software and services running on them, which inevitably means more vulnerabilities.

To derive the remediation capacity of the firms in our sample, we first calculated their average number of closed and open vulnerabilities per month. We then used those to construct a regression model, seen in Figure 5.4.

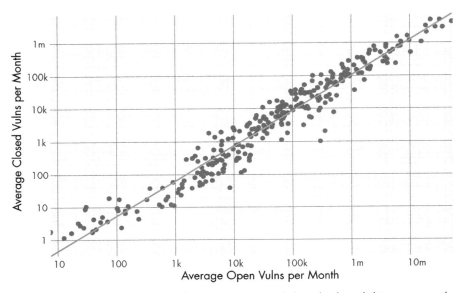

Figure 5.4 Regression model of average open and closed vulnerabilities per month. (© 2019 Kenna Security/Cyentia Institute. Reprinted with permission [5].)

Each point represents the remediation capacity for each organization and fits tightly along the predicted trendline. So tightly in fact that the R2 statistic for this log-log regression model is 0.93, meaning that it's very strong and captures most of the variability around vulnerability closure rates.

Strong models are great (especially when they're so simple), but there's something else this figure teaches us that's greater still. Notice first that each axis is presented on a log scale, increasing by multiples of 10. Now, follow the regression line from the bottom left to upper right. See how every tenfold increase in open vulnerabilities is met with a roughly tenfold increase in closed vulnerabilities? That, in a nutshell, is why many firms feel like their vulnerability management program can never pull ahead in the race of remediation.

But progress is possible. If we add a second regression model for the top performers (orange), you can see that both the line and each point on the graph fall above the original line (Figure 5.5).

That means top performers achieve a higher remediation capacity than other organizations. The constant of remediating one in 10 vulnerabilities per month is not a ceiling. Top performers, in fact, had a remediation capacity of 2.5 times that of other firms (Figure 5.6).

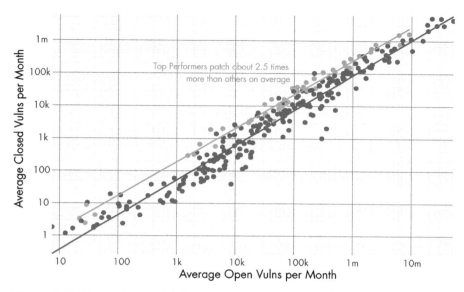

Figure 5.5 Regression model showing average open and closed vulnerabilities per month for top performers versus everyone else. (© 2019 Kenna Security/Cyentia Institute. Reprinted with permission [5].)

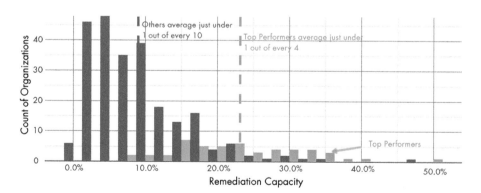

Figure 5.6 Comparing remediation capacity of top performers vs others. (© 2019 Kenna Security/Cyentia Institute. Reprinted with permission [5].)

All of this is to say that by using the right remediation metrics, you can not only gain a more accurate picture of your performance in the context of the overall volume of vulnerabilities in your environment, but that by looking closely at that picture, you can find ways to improve your average remediation rates by increasing your velocity and capacity.

We'll look at each of them—and the trade-offs required—in a moment, but first a quick word on why we're bothering to measure all of this in the first place.

5.4 WHY DOES PERFORMANCE MATTER?

Scale, speed, and operations are great defenses against the threats every organization faces. But they most often support symmetrical responses to threats. An attacker develops an exploit, and we develop a countermeasure. The attacker finds a vulnerability, and we apply a patch. As Richard Seiersen has put it, our only asymmetric advantage is analytics [6]. By measuring the overall statistical distribution of vulnerabilities and exploits, we can forecast what is more probable and impactful over time.

But that advantage can be lost without a continuous loop of performance feedback that reflects your progress and priorities. The overarching goal of any vulnerability management program is to avoid a cybersecurity incident. Your risk of suffering an incident decreases as you track and improve the right metrics. Measuring performance acts as a gateway to reduced risk by ensuring an accurate view of where you stand against your most critical vulnerabilities.

Performance metrics couple your vulnerability remediation efforts to reality. In this case, reality is risk. If you are doing well in these measures, you should be reducing your risk. Unfortunately, many organizations focus on poor performance indicators, see improvements, and think they've achieved progress against risk. The reality might be the opposite.

A clear-eyed view of performance offers actionable information that allows you to achieve incremental improvements to your entire vulnerability management program. The organizations able to achieve higher levels of performance and greater success focus relentlessly on accurate measures of performance. The metrics you choose to focus on will reflect your priorities as an organization.

Ultimately, performance metrics bubble up and influence whether you spend on remediation or hiring. Some metrics make their way up to the board to tell a story, while others will be much more operational—used to measure where you stand on a day-to-day or hour-to-hour basis. Either way, metrics can convince stakeholders the program is worthwhile and successful, and win budget for further improvements.

With that said, let's look more closely at the metrics that will give you a true sense of your performance and continually improve your exposure to risk.

5.5 MEASURING WHAT MATTERS

The journey to measuring progress starts at stage zero, which is best described as "ignorance is bliss." At that stage, the organization isn't doing any kind of vulnerability assessment and has no idea where vulnerabilities reside.

From there, they start scanning and realize vulnerabilities are everywhere. Panic erupts. Teams are overwhelmed. Denialism and all the stages of grief set in. They realize they're never going to fix all the vulnerabilities, will never have enough resources, and move to a risk-based approach.

At that stage, their risk is still high and they can make quick progress by zeroing in on fixing the riskiest vulnerabilities—a focus on efficiency. But as their program matures, there are several metrics they should focus on, both to notch quick wins in reducing risk and keep their risk low over the long term.

When evaluating remediation strategies, it's tempting to opt for overall accuracy—vulnerabilities remediated that were actually exploited—but this can be misleading when so many vulnerabilities are never exploited. For example, a company choosing not to remediate anything will be "right" around 77% of the time because 77% of vulnerabilities are never exploited. It might look like a good strategy on paper, but not so much in practice, because they didn't remediate any of the 23% of the vulnerabilities that were exploited.

What really matters is your coverage and efficiency, remediation capacity and velocity, vulnerability debt, and remediation SLAs.

5.5.1 Coverage and Efficiency

The first two metrics you should focus on are coverage and efficiency. You'll recall from Chapter 1 that coverage refers to the completeness of your remediation. Of all vulnerabilities that should be remediated, what percentage did you correctly identify for remediation? Efficiency refers to the precision of your remediation. Of all vulnerabilities identified for remediation, what percentage should have been remediated?

Ideally, we'd love a remediation strategy that achieves 100% coverage and 100% efficiency. But in reality, a trade-off exists between the two. A strategy that prioritizes only "really critical" CVEs for remediation has a high efficiency rating, but this comes at the cost of much lower coverage. Conversely, we could improve coverage by remediating more vulnerabilities by using a strategy like remediating all vulnerabilities CVSS 5 or above, but efficiency drops significantly due to chasing down vulnerabilities that were never exploited.

One common strategy called CVSS 7+ aims to remediate vulnerabilities with a CVSS score of 7 or above. However, this is not an efficient strategy because CVSS 7+ has an efficiency of only around 31% (Figure 5.7).

Figure 5.7 shows how most of the traditional remediation strategies do in relation to coverage and efficiency, and we also show several strategies that optimally balance coverage and efficiency.

These strategies, represented by the red line, are the "highly efficient," "balanced," and "broad coverage" strategies. Notice that all of the traditional remediation strategies that are based only on CVSS scores do poorly relative to the strategies on the red line.

5.5.1.1 Optimizing the Trade-Off between Coverage and Efficiency with Predictive Models

A predictive model allows you to choose a remediation strategy that stays on the red line, where the red line represents the optimal trade-off of efficiency and coverage for a given level of effort. When compared to the CVSS 7+ strategy, the "balanced" strategy that uses a predictive model achieves twice the efficiency (61% vs. 31%), with half the effort (19K vs 37K CVEs), and one-third the false positives (7K vs 25K CVEs) at a better level of coverage (62% vs 53%).

Why does a strategy based on a predictive model perform so much better in terms of efficiency, effort, coverage, and number of false positives than traditional remediation strategies? Because it incorporates different types of data that are high indicators of risk through statistical analysis.

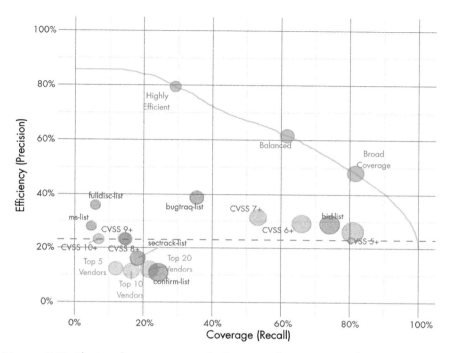

Figure 5.7 Plotting the coverage and efficiency of various remediation strategies. (© 2019 Kenna Security/Cyentia Institute. Reprinted with permission [7].)

Those indicators of risk include whether the vulnerability has exploit code written for it or the actual prevalence of the vulnerability in the wild. In other words, a good predictive model uses more than one type of data, does a much better job of accounting for data interaction effects, and is even able to leverage dynamic data.

For example, one of the shortfalls of CVSS scores is that they are static. Once a CVSS score is set, it rarely changes. A good predictive model will make use of dynamic data, so if the risk changes, the score also changes. In other words, for a given level of effort (resources) a remediation strategy that uses a predictive model is both more efficient and provides better coverage.

5.5.1.2 Coverage and Efficiency in the Real World

The powerful thing about the goals of coverage and efficiency is that they can be objectively measured and compared among different remediation strategies. In a sample of 12 organizations selected across a range of sizes and industries to highlight variation in remediation strategies and outcomes, most organizations appear to sacrifice efficiency for the sake of broader coverage. They are addressing the majority of higher-risk issues (70% coverage) but paying a relatively high resource cost to achieve that goal.

This is classic risk-averse behavior, and undoubtedly driven by the comparatively higher cost of getting it wrong and suffering a breach. These 12 firms observed 190 million vulnerabilities across their networks, of which 73% (139 million) had been closed at the time of the study. We also included the outcome of several CVSS-based remediation strategies for comparison.

Given what we saw in Figure 5.7 about the efficacy of remediation strategies, it shouldn't be surprising that CVSS did not fare well as a prioritization strategy in the real world. Just keying on CVSS 10+ would address about a fifth of the vulnerabilities that should be fixed, and four-fifths of that effort would be spent on vulnerabilities that represent lower risk and thus could have been delayed.

Among the 12 organizations, several approach 100% on the coverage axis. That indicates they're staying on top of the high-priority vulnerabilities. There's another cluster around the 75% coverage mark, which still isn't bad. One firm falls below the 50% line, meaning it has remediated less than half of vulnerabilities with known exploits. This spread of coverage levels is interesting, but the burning question from Figure 5.8 is why is efficiency so low?

Put simply: Patches are inefficient, at least when it comes to unmodified calculations of coverage and efficiency. When we speak of remediating vulnerabilities, many times the actual decision is whether or not to apply a patch. The relationship between vulnerabilities and patches is often not a simple one-to-one. In the hundreds of thousands of patches we've collected, about half fix two or more CVEs.

Let's say the patch you deploy fixes five CVEs and only one of those is exploited. According to the raw efficiency calculation, you chose "wrong" four out of five times. But you really didn't choose to remediate those other four, so your efficiency metric is inherently penalized by patching. But we all know patching isn't really inefficient, so we have a challenge to calculate a cleaner efficiency rating that accounts for patches.

We have the data to do this, starting with which CVEs have been observed and/or exploited. We also have data linking CVEs to

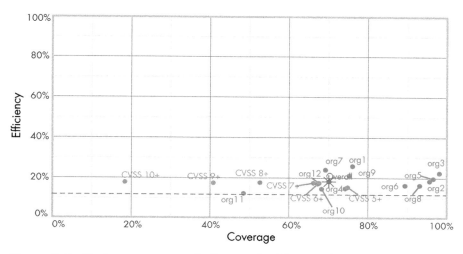

Figure 5.8 Plotting coverage and efficiency for 12 example organizations, along with estimates of CVSS-based scores for comparison. (© 2019 Kenna Security/Cyentia Institute. Reprinted with permission [7].)

observed vulnerabilities as well as CVEs to one or more patches. The result is Figure 5.9, which shows the jump in efficiency once we account for patches.

Notice the lift in efficiency via this alternate method of calculating efficiency. This offers a more realistic view of coverage and efficiency for the organizations depicted. Note that the degree of change on the efficiency axis isn't uniform. Some organizations are clearly making more efficient patch-centric remediation decisions than others.

There's also an intriguing variation among the firms when using this alternate method of measuring efficiency. The range between the minimum and maximum is roughly 50% for both coverage and efficiency. This suggests vulnerability management programs really do matter and a real, measurable improvement can be gained by making smarter remediation decisions.

5.5.2 Velocity and Capacity

The velocity of your remediation efforts—the speed and direction—can be measured via the AUC in a survival analysis. A lower AUC points to a shorter survival rate; higher means the opposite.

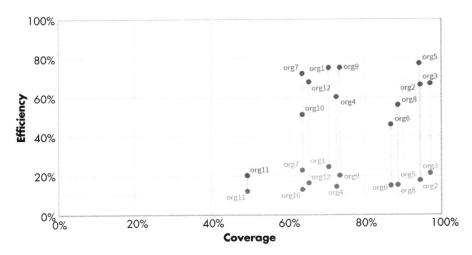

Figure 5.9 Plotting coverage and efficiency for 12 example organizations, adjusted to account for patching. (© 2019 Kenna Security/Cyentia Institute. Reprinted with permission [7].)

For our application, a lower AUC translates to a shorter lifespan for vulnerabilities.

Figure 5.10 shows survival analysis curves for four organizations of varying success at remediation. It's easy to see the relationship between the shaded portion under each curve and the resulting AUC. It's also easy to see how the firm on the left drives down vulnerabilities with more expediency than the others, resulting in a greater velocity.

Tracking this change across all of the vulnerabilities across many organizations over time produces a curve like the ones shown in Figure 5.11.

Across the 103 firms included, we see that the overall half-life of a vulnerability is 159 days. But we now also have a view of the long tail of vulnerabilities that remain open for well over a year. The second graph shows the various curves of the organizations included in the data. As we've noted before, remediation timeframes vary substantially.

Many factors impact velocity, including the vendors that created the vulnerabilities in the first place. For example, in our study of 300 organizations, it took them 15 times longer to address half their vulnerabilities affecting Oracle, HP, and IBM products than to reach

5.5 MEASURING WHAT MATTERS   125

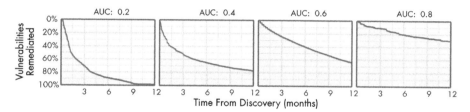

Figure 5.10 Example AUC measures for four organizations. (© 2019 Reprinted with permission of Kenna Security/Cyentia Institute [5].)

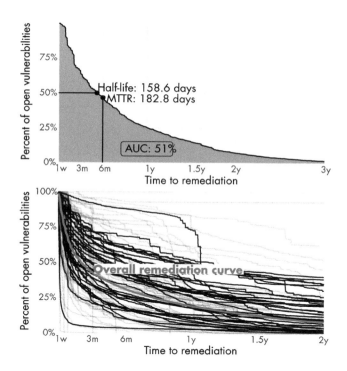

Figure 5.11 Survival analysis curves for vulnerability remediation timelines. (© 2019 Kenna Security/Cyentia Institute. Reprinted with permission [8].)

that same milestone with Microsoft products (see Figure 5.12). What's more, moving from 50% to 75% remediation takes multiple years for several vendors.

Keep in mind that some CVEs are not part of the normal patching process, either because of complexity, as was the case with Meltdown,

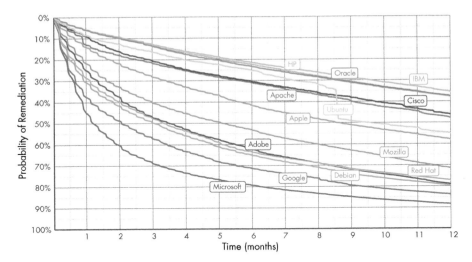

Figure 5.12 Remediation velocity for major product vendors. (© 2019 Kenna Security/Cyentia Institute. Reprinted with permission [5].)

or media attention, as with BlueKeep. For these kinds of vulnerabilities, the median vendor metric might not be useful. Some teams are not aligned to vendors or products either, and are responsible for larger or smaller swaths of the environment. In that case, the slice of data should be aligned with the particular responsibilities of the teams. We suggest allowing for any permutation of metric measurement.

In particular, there's value in looking into the velocity of individual CVEs and the stages of their lifecycle, which affect velocity. Looking at 473 vulnerabilities with known exploits in the wild, our friends at the Cyentia Institute calculated the median number of days between the milestones in the life cycle of those vulnerabilities (Figure 5.13).

Of course the order of milestones varies widely. Figure 5.14 shows the top 10 most common sequences, but there are more than 100 other unique sequences.

Zooming in on one particular CVE illustrates the variability that can occur within the vulnerabilities for a particular vendor. Take CVE-2020-0688 for example. Released in February 2020, this CVE details a static cryptographic key in Microsoft Exchange Server's on-by-default ECP control panel. By March, it was actively exploited in the wild by

Figure 5.13 Average timeline of key milestones in the vulnerability lifecycle. (© 2019 Kenna Security/Cyentia Institute. Reprinted with permission [9].)

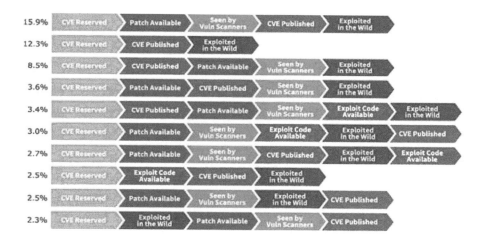

Figure 5.14 Top 10 sequences of key milestones in the vulnerability lifecycle. (© 2021 Kenna Security/Cyentia Institute. Reprinted with permission [9].)

threat actors, and both US-CERT and the NSA issued guidance urging teams to quickly patch this vulnerability.

At the time that guidance was released, the vulnerability was less than 15% remediated, far short of the 50% typical for Microsoft-source vulnerabilities a month after the announcement.

Among the companies that hadn't yet remediated CVE-2020-0688, perhaps their servers weren't directly exposed to the internet or users worked around and disabled ECP or were operating on a domain that doesn't have as much exposure to phishing or other credential leaks, making this patch less critical.

Exchange servers are also notoriously hard to patch and hard to upgrade, and since Exchange is a critical organization service, it may be off limits to normal monthly patching schedules. While that certainly makes patching more difficult, it also makes the vulnerability an attractive target.

As you can see, even within the relative speed of patching for a particular vendor, there are particular CVEs where patching is more difficult or time consuming for various reasons. Just as looking at the velocity of remediation for various vendors can indicate where to focus your attention or resources, looking at each CVE can reveal opportunities and challenges to address.

Velocity can vary by industry as well. In one study, healthcare institutions take about five times longer than leading industries to close half their vulnerabilities (Figure 5.15). Banking was in the middle and insurance was dead last.

It's hard to fathom educational institutions remediating vulnerabilities faster than banks, but that's exactly what the data shows here, at least to the 25% milestone. Those lines diverge quickly after the midway point in banking's favor, suggesting that being quick out of the gate isn't necessarily the best way to win the race.

The size of the organization can also affect velocity, but in interesting ways. While smaller firms certainly have fewer resources, they also tend to have smaller-scale problems. Larger firms have the

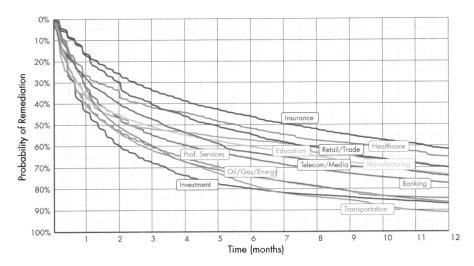

Figure 5.15 Remediation velocity by industry. (© 2019 Kenna Security/Cyentia Institute. Reprinted with permission [5].)

luxury of more resources but not without more problems to boot. In that sense, remediation velocity may be a self-normalizing metric across firm sizes.

While many assume fewer resources would translate to reduced capacity to remediate vulnerabilities, smaller firms generally reach each time-to-fix milestone faster than their medium and large counterparts (Figure 5.16).

That trend is even more apparent when looking at exploited vulnerabilities. Notice in Figure 5.16 the time interval between 50% and 75% remediation. First, we see noticeably shorter survival times for the exploited vulnerabilities, which is good. The second observation is more subtle but has important implications. Notice how the 50% to 75% interval for nonexploited vulnerabilities in large organizations is extremely long. It's almost like they've accepted they don't have the capacity to fix everything and have shifted resources accordingly to get more bang for the buck on those riskier vulnerabilities.

It's important to remember that it's not necessarily a bad thing that organizations don't close all vulnerabilities quickly. If an organization dedicated all its resources to drive to zero as fast as possible on every vulnerability, that would be downright wasteful, since as we've seen, 77% of CVEs have no published or observed exploit, making them much less risky.

Just as there's a trade-off between coverage and efficiency, there's one between velocity and efficiency. Organizations with a high remediation velocity tend to have lower efficiency, meaning they remediate vulnerabilities that would never have been exploited. The cost of not remediating, or remediating too slowly, is certainly potentially much higher than over-remediating. The more important question is

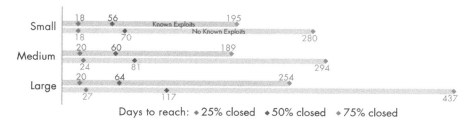

Figure 5.16 Remediation velocity by organization size. (© 2019 Kenna Security/Cyentia Institute. Reprinted with permission [5].)

how quickly a company remediates the vulnerabilities that pose the greatest risk to their organization.

While some firms address vulnerabilities with a swift velocity, others play the long game with a higher capacity. Capacity represents how many vulnerabilities an organization can remediate and the degree to which high-risk vulnerabilities build up over time—are you ahead of the game, treading water, or falling behind?

Remediation capacity is determined by two primary metrics. MMCR measures the proportion of all open vulnerabilities a firm can close within a given time frame. Vulnerability debt measures the net surplus or deficit of open high-risk vulnerabilities in the environment over time.

Think of MMCR as raw remediation capacity. To derive it, we calculate a ratio for the average number of open and closed vulnerabilities per month. You'll recall that, on average, organizations remediate about one out of every 10 vulnerabilities in their environment within a given month.

There's variation around this 1-in-10 ratio, of course, and Figure 5.17 demonstrates that fact. In a survey of 103 organizations, quite a few firms fall below that mark and some exceed it by a large margin. It's tempting to assume those exhibiting higher remediation capacity must have less infrastructure to manage, but the data doesn't support that conclusion. Average capacity remains remarkably consistent, regardless of characteristics like organization size, number of assets, and total vulnerabilities.

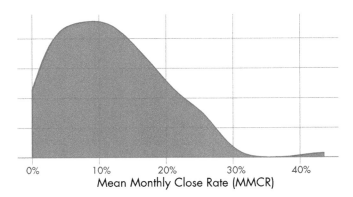

Figure 5.17 Mean monthly close rate. (© 2019 Kenna Security/Cyentia Institute. Reprinted with permission [8].)

5.5.2.1 How Much Does Capacity Cost?

As we saw earlier in this chapter, it's possible to increase capacity and remediate more high-risk vulnerabilities in any given time frame. In fact, the top performers in our study had a capacity 2.5 times that of other organizations. But just as there's a cost to pursuing any individual measure, usually in the form of a trade-off, there's a literal cost to increasing capacity. You can increase your capacity by adding more people or more automation, both of which are costly.

Another strategy is to ensure you're maximizing your current capacity through better triage. Team structure plays a role. We'll cover this more in a later chapter, but suffice it to say, MTTR is about a month and a half shorter among firms that house and fix responsibilities in separate organizations (Figure 5.18).

Prioritization matters too, of course. We found similar time savings when reducing the scope to just high-risk vulnerabilities (Figure 5.19).

If you have the capacity to fix 100 vulnerabilities per month, you can make sure you're best utilizing that capacity by adjusting the dials on your prioritization so you're relentlessly prioritizing the remediation of high-risk vulnerabilities. Your high-risk capacity goes up, even as your capacity remains the same, allowing you to make more progress in lowering risk.

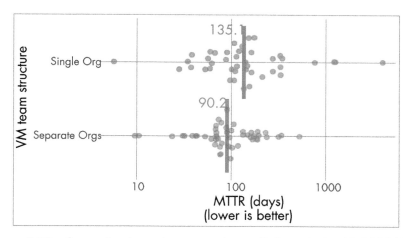

Figure 5.18 Comparing team structure and remediation velocity. (© 2019 Kenna Security/Cyentia Institute. Reprinted with permission [8].)

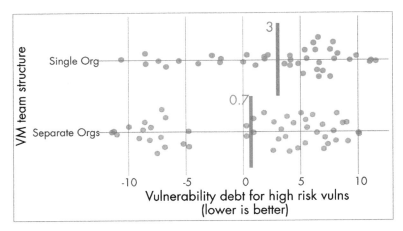

Figure 5.19 Comparing team structure and remediation capacity. (© 2019 Kenna Security/Cyentia Institute. Reprinted with permission [8].)

5.5.2.2 The Power Law of Capacity

In the last chapter, we looked at how power law randomness requires us to shift our thinking about the threats we face in cybersecurity compared to the threats faced in a physical battle.

This is especially important to keep in mind with respect to capacity. Just as the randomness we encounter in exploits and threats follows a power-law distribution, so must our response. This is another reason why MTTR and other mean-time-tos are not great metrics—they focus on the average, when a single vulnerability might be your greatest source of risk.

When using a risk-based approach to security, firms recognize their limited capacity. Even as you're considering the riskiest events to your organization, any one vulnerability could end up being 100× riskier than the last one because there's no stable average. How you respond to risk starts to become important, and by staggering your capacity, you can aim to remediate the riskiest vulnerabilities orders of magnitude faster than others.

Staying ahead of the vulnerabilities that represent the highest risk brings us to a deeper dive on the concept of vulnerability debt.

5.5.3 Vulnerability Debt

Most organizations are either building a backlog of open vulns or slowly chipping away at them. We turned this into a metric by tracking the number of open vulnerabilities an organization has over time—the concept of vulnerability debt that we introduced in the last section.

We estimated this change over time by extracting the slope of a linear regression for each organization (Figure 5.20). Of course some organizations have orders of magnitude more vulnerabilities than others, so we scaled the slope so we can compare apples to apples.

We also calculated the total number of open high-risk vulnerabilities and the number remediated per month for each organization. The resulting ratio, depicted in Figure 5.21, identifies which firms are keeping up (closing about as many vulnerabilities as were opened), falling behind (opened > closed), or gaining ground (closed > opened).

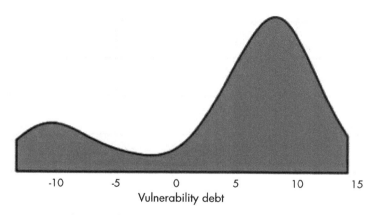

Figure 5.20 Distribution of vulnerability debt. (© 2019 Kenna Security/Cyentia Institute. Reprinted with permission [8].)

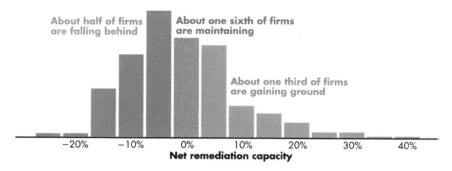

Figure 5.21 Comparison of net remediation capacity for high-risk vulnerabilities. (© 2019 Kenna Security/Cyentia Institute. Reprinted with permission [8].)

From that, we see about half of firms are falling behind, one in six are keeping up, while one in three are gaining ground. Figure 5.21 also makes it plain that the degree by which organizations are falling behind or pulling ahead varies widely.

5.5.3.1 The Move to the Cloud

All of these factors are also influenced by the move to the cloud. On the one hand, moving to the cloud could decrease your vulnerability debt and free up capacity. First, the cloud introduces more automated processes, which can certainly contribute to capacity that allows you to make progress against debt.

Second, if you're running a standard virtual machine for a web app that runs off a container image in the cloud, Amazon is responsible for keeping that up to date. If you have a vulnerable image that you're running in the cloud or a container, Amazon will alert you to potential vulnerabilities and offer ways to remediate if you don't want to update. The cloud in some ways puts the burden on someone else for those security costs, leaving your capacity focused on what you manage.

On the other hand, the cloud can also increase vulnerability debt. In most organizations, the cloud doesn't replace anything, it's used in addition to the assets they already have in place. That means their attack surface has expanded. Organizations aren't decommissioning applications and moving them to the cloud. They're just moving to the cloud, or adding to the cloud, which potentially piles on to your vulnerability debt.

Prioritization and automation can both help. Automation allows you to fix more, and in that sense, the tooling available in the cloud natively helps. As organizations move toward immutable infrastructure, instead of applying a patch they can turn down the old instance and turn up a new instance that's clean and updated. In addition, the advent of containers means firms can start to remove old containers for new containers.

This makes it a lot easier to remediate vulnerabilities, but it's still a long, slow journey. Especially since many, if not most, of the organizations using containers in the cloud treat them like virtual machines, not like containers. The containers end up having long lives, and organizations even patch them, which is not the intent but that's how they're being used.

While the cloud is still mostly expanding an organization's attack surface, that will change over time, and the automation and efficiencies of the cloud will chip away at, rather than add to, organizations' vulnerability debt.

5.5.3.2 Paying Down Security Debt

When an organization first embarks on a risk-based vulnerability management strategy, they discover a lot of debt. Their efficiency is high as they make their first strides at reducing risk. As they mature, they look to level out and lower risk across the entire organization, so they shift strategies from efficiency to coverage.

We saw this in the survival analysis curves earlier in the chapter, where organizations' velocity in closing 25% of their vulnerabilities is especially swift, and even closing 50% occurs relatively quickly, then levels off from there as the long tail of vulnerabilities stretches out beyond a year.

As we've discussed, this isn't necessarily a problem in a risk-based program. When you're properly prioritizing the remediation of the riskiest vulnerabilities, you'll be able to quickly pay down your most critical debt in days, and then make progress on the rest in a time frame fitting their risk.

To ensure you're paying down your debt consistently and focusing on fixing what matters first, at a certain point your organization will need to set appropriate SLAs.

5.5.4 Remediation SLAs

SLAs define deadlines for vulnerability remediation time frames. They're also a sign of maturity. Going from no SLA to implementing an SLA can create a surge in velocity and overall performance. Once an SLA is in place, incremental improvements over time make the time frames more stringent and advance progress. The most successful organizations inject risk into their SLAs so that they're not just based on a severity score or a technology stack, but a holistic view of the risk a vulnerability poses.

It's important to start with a simple SLA. SLAs can quickly become complex, especially if your organization is in a more heavily regulated industry, like banking or insurance, and you need to factor that into the risk tolerance. But at a basic level, you can design SLAs around one of three categories: As fast as your peers, faster than your peers, or faster than attackers.

Being faster than attackers is a completely different game and requires incredibly fast remediation for the riskiest vulnerabilities and also pushing out the deadlines for less risky vulnerabilities to a year or longer—possibly forever—because they'll never be attacked.

How you design your SLAs depends on your organization's goals and risk tolerance. But just having an SLA often improves your ability to remediate the vulnerabilities that matter.

In a survey of about 100 organizations, about two-thirds of them had defined deadlines and followed them in some capacity (Figure 5.22). The remainder either handled vulnerabilities on a case-by-case basis or aren't sure if SLAs exist (which would seem to suggest they do not).

We then asked respondents with defined deadlines how many days their firms allow for fixing vulnerabilities of varying priority or risk levels. Not surprisingly, remediation windows vary both within and across priority levels. The peaks of the distributions in Figure 5.23 point to a week for the highest priority vulnerabilities, a month for high priorities, and three months for the moderates.

And they work. Our analysis suggests that firms with defined SLAs address vulnerabilities more quickly than those with nonexistent or ad hoc schedules. In terms of metrics, that's observed as a 15%

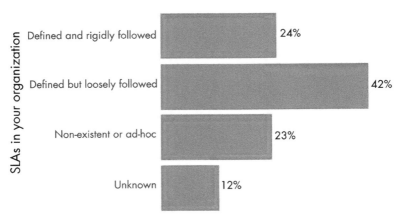

Figure 5.22 Status of SLAs in a survey of 100 organizations. (© 2019 Kenna Security/Cyentia Institute. Reprinted with permission [8].)

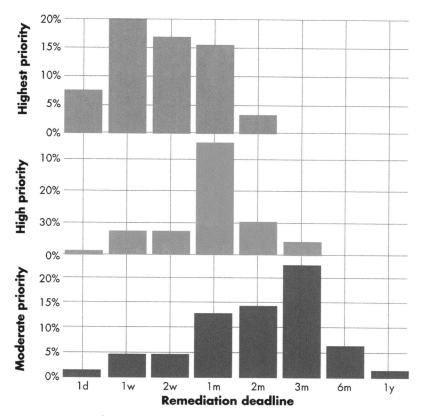

Figure 5.23 Remediation deadlines by priority level. (© 2019 Kenna Security/Cyentia Institute. Reprinted with permission [8].)

improvement to AUC for high-risk vulns and 10% better AUC across all vulns (Figure 5.24). Sure, deadlines are missed from time to time, but not having them invites making it a habit.

In addition to boosting velocity, it seems that priority-driven SLAs correlate with expanded overall remediation capacity as well. Adding a few percentage points to the proportion of vulnerabilities you can close in a given month may not seem like much, but a point here and a point there may be the difference between running a surplus versus adding to the deficit.

We also compared remediation velocity among organizations with shorter and longer SLAs within each priority level, but found no major differences. In other words, firms on the left side of the

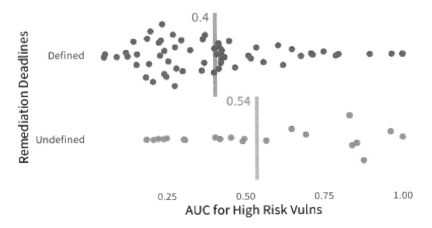

Figure 5.24 Impact of defined and undefined SLAs on remediation velocity. (© 2019 Kenna Security/Cyentia Institute. Reprinted with permission [8].)

distribution in Figure 5.23 didn't fix vulnerabilities significantly more quickly than those toward the right. That's a bit counterintuitive to be sure. Our takeaway is that establishing SLAs has more impact on performance than how aggressive remediation windows are in relation to others. From a practical perspective, that may simply translate to "set realistic goals."

As you do, consider that vulnerabilities fall along a power-law distribution and factor that into your remediation deadlines. A small portion of vulnerabilities will need to be remediated nearly immediately and the vast majority won't matter much in the grand scheme of things.

So while SLAs based on 30-, 60-, and 90-day remediation timelines aren't uncommon, think about other distributions that leverage power laws. For example, you might have an SLA of 7 days for the highest risk vulnerabilities, 30 days for the middle tier, and 2 years for the lowest tier, the long-tail vulnerability debt.

At this point you should understand why measuring performance matters, how to assess whether a particular performance metric is useful, and the most important metrics to track to gauge your success in remediating the vulnerabilities that pose the greatest risk.

References

[1] Swan, T., "Ford's Assembly Line Turns 100: How It Really Put the World on Wheels," *Car and Driver*, April 30, 2013, https://www.caranddriver.com/features/a15115930/fords-assembly-line-turns-100-how-it-really-put-the-world-on-wheels-feature/.

[2] Goss, J. L., "Henry Ford and the Auto Assembly Line," *ThoughtCo*, January 22, 2020, https://www.thoughtco.com/henry-ford-and-the-assembly-line-1779201.

[3] Agence France-Presse, "Ford Launched the Modern Assembly Line a Century Ago and Changed Society," *IndustryWeek*, October 7, 2013, https://www.industryweek.com/innovation/process-improvement/article/21961364/ford-launched-the-modern-assembly-line-a-century-ago-and-changed-society.

[4] Kelley, T. L., *Interpretation of Educational Measurements*, Yonkers-on-Hudson, NY: World Book Company, 1927, p. 14.

[5] Kenna Security and The Cyentia Institute, *Prioritization to Prediction Volume 3: Winning the Remediation Race*, 2019, https://learn-cloudsecurity.cisco.com/kenna-resources/kenna/prioritization-to-prediction-volume-3.

[6] Personal communication, *SIRAcon 2018*, https://www.rsaconference.com/experts/richard-seiersen.

[7] Kenna Security and The Cyentia Institute, *Prioritization to Prediction Volume 2: Getting Real About Remediation*, 2019. https://learn-cloudsecurity.cisco.com/kenna-resources/kenna/prioritization-to-prediction-volume-2.

[8] Kenna Security and The Cyentia Institute, *Prioritization to Prediction Volume 4: Measuring What Matters in Remediation*, 2019, https://learn-cloudsecurity.cisco.com/kenna-resources/kenna/prioritization-to-prediction-volume-4.

[9] https://www.kennasecurity.com/blog/flaw-of-averages-in-averaging-flaws/.

6
BUILDING A SYSTEM FOR SCALE

At this point, you're ready to take your first steps toward a risk-based vulnerability management program. You understand the underlying mathematical concepts and how to build a decision engine to score the risk of a particular vulnerability and determine how much of a threat it poses to your organization. You understand how to interpret those scores and measure the performance of that system.

But before you start to build, it's important to think several steps ahead on the path your RBVM program will take as it matures. Everything from the tools you have to your organization's technology roadmap to which users will have access to the platform all play a role in how you build it.

No matter the industry or organization, there's a lot to consider when building a system for scale. Let's look at the strategies behind a system that can mature alongside your RBVM program.

6.1 CONSIDERATIONS BEFORE YOU BUILD

There are several factors that impact your system's ability to scale, and you should consider all of them before you begin building.

Start with your assets. You should know all of the classes of assets in your environment, from on-premise to the cloud. But you also need to know the volume of those assets and their associated metadata and other information. Where will you be able to store all that data? Will your SQL database be sufficient? Do you need to implement sharding or take other measures to manage the volume of data around your assets? It's not just the number of assets either; it's also coupling those assets together and grouping that is going to affect compute more than storage.

Even more important is your organization's vulnerability density. Even if you have thousands of assets, scaling might not be that difficult a task if they're mostly clean. On the other hand, if you have a lot of vulnerabilities on a lot of assets, that will be problematic if you don't plan for that when building out the system.

How clean or dirty a system is correlates strongly to the industry your organization is part of. For example, if your organization is in technology, you most likely will have many assets that are mostly clean—you won't have a particularly high vulnerability density. The opposite could be said for healthcare. In that case, it's much more likely your organization will have a high vulnerability density and you'll have to take the appropriate measures in compute and storage.

It's also important to consider scale in terms of users, even if the amount of data (about your assets and vulnerabilities and how messy that data is) will determine in large part what your system looks like. That includes not only the total number of users, but the types of users and what role they play in your organization, how they'll use the system, and how you can take steps toward self-service capabilities as you scale and mature your RBVM program.

Once you have a better sense of the users, you can begin to find efficiencies in the building process. For example, if everybody in the organization will be using the solution, can you tie it to SAM or active directory where people are already logged in, and can you manage all of the roles based off of the groups that are in active directory? Obviously, that matters in terms of scale when you're setting up the system. You don't want to go in and create 10,000 users individually when you can access them from another system that already exists and already takes into account how to structure and group their roles.

One final aspect to consider in terms of users is where they are in the world. If you're standing everything up for users localized to the east coast, AWS US-East-1 in Northern Virginia might work just fine. But if you have users in Australia as well, that AWS region probably won't be acceptable. Will you need to have multiple zones, and do you need to do any sort of front-end caching to optimize both performance and the user experience?

All of these considerations are only the beginning. Let's look more in depth at three additional considerations that will play a major role in your thinking and strategy as you build your system.

6.1.1 Asset Management Assessment

One of the first things to do is assess where you stand with your asset management. In Chapter 4, we discussed asset scanning and its importance in an RBVM program. To build a system for scale, you need to understand your ability to identify and manage those assets in your environment.

If you're new to asset management, getting past the initial scan can be daunting because you're starting from scratch and scanning your entire environment for the first time. But the difference in going from nothing to something has a strong upside. The initial push can get you 70% to 80% of the way to where you need to be, and from that point, there are a number of improvements you can make to fine-tune your asset management.

The scale of this project will vary by organization. At a startup or a smaller organization, for example, you'll likely know where to find most of the assets in your environment. You might look at AWS, GCP, or Azure, tap into Jamf for MDM, and so on. By looking in just a few places, you might be able to find and understand 99% of the picture.

If you work for General Motors, on the other hand, it's a completely different story. You need scale just in terms of the sheer volume of assets. You need to consider the infrastructure you need to both discover those assets as well as store and process all the configuration data about them. The breadth and complexity of the asset data is most important.

The key to scaling is breaking the assets up into categories. You'll have endpoint devices, IoT devices, on-premise data centers, cloud storage, applications, and workloads, and so on. From there, look at the tool sets you can use to discover assets in each category.

Ultimately, what you need to consider around the breadth and complexity of that data is the data model you're using around it. Any given asset has a number of attributes to track that will allow you to make decisions about it, including who owns the asset and the risk the asset poses. You can use all of this information to manage each asset moving forward. The most difficult—and critical—part is making sure you know who owns each asset.

6.1.2 Where Your Organization Is Going

In addition to implementing the proper tools for scaling your asset discovery and management, it's important to have an overall picture of where your organization is going and what you need to factor into the build.

For example, is your organization starting to deploy more workloads in the cloud? Are you moving more toward containers? What is at the forefront of your tech that you need to make sure is accounted for moving forward when doing vulnerability assessment and management that will affect your organization down the road?

Think about the data you're building out. What is an asset? The definition might shift when you start to move toward the cloud. Workloads and applications will be more ephemeral.

How you actually fix vulnerabilities will also change quite a bit. If your applications are on premise and you identify a vulnerability, you deploy a patch. If you have a vulnerability in the cloud, you'll be doing less patching because you can turn down a vulnerable system and turn up a brand-new system that's updated without those vulnerabilities. In that situation, you'll probably have more automation in place, which allows you to fix more assets more quickly and efficiently when you're in the cloud.

In risk-based vulnerability management, you really have two levers. The first is prioritization. You have a bunch of vulnerabilities, but you can't fix them all, so you need to prioritize down to what your organization is capable of fixing. The second lever is automation, which allows you to fix a lot more vulnerabilities with the same number of employees. In the cloud, you can leverage a lot more than automation, too, which could affect your prioritization. Your risk-based SLA, which defines what you will focus on, might broaden because your capabilities for remediation have increased through automation.

Application security and application vulnerabilities matter as well in terms of where your organization is going. You need a full stack view of not just the infrastructure vulnerabilities, but all the different layers of the application as you move to the cloud.

Those lines start to blur. Traditional on-premises infrastructure was fairly easy to track. You had servers, desktops, laptops, and so on, each with vulnerabilities associated with it. You had applications you built, again, each with vulnerabilities. As you move to the cloud, however, it can become harder to define exactly what is infrastructure and what is an application.

You might have a container that's actually running a microservice—is that an application or is that infrastructure? You might fix it with code by turning up and turning down systems rather than patching. But there's also the concept of infrastructure as code, using tools like Chef, Puppet, Ansible, and Terraform, where if you want to fix a vulnerability. you don't deploy a patch, but instead update a line of code that changes all of the microservices or containers so you're no longer vulnerable.

6.1.3 Other Tools as Constraints

When you think about tools, there are constraints in two different senses.

The first sense is something we've already talked about. If you have a network vulnerability scanner, you're typically not going to use that in a cloud environment. The closest you'll come to that is an agent from the same provider that's running on a system or side channeling through Amazon logs to understand where your vulnerabilities are. In this case you won't use a scanner at all. So the assessment tools you use can have some constraints depending on what layer you're looking at.

In addition, there are tools that act as the opposite of a constraint—they're intended for one use but can offer valuable information to security teams as well. When Kenna Security's CTO and Co-founder Ed Bellis worked at Orbitz, the company used ClickStream software for customer service. The company would use it to identify issues within the product when customers experienced frustration. Based on customers' clicks and the errors they ran into, plus the help desk tickets the customers submitted, Orbitz could look back and identify the problem the users experienced and make changes and updates to improve the user experience across the board.

In security, we can repurpose the exact same software to look for abnormalities. For example, if you see that a user starts down the path of booking an airline ticket, but jumps from step one to step seven, that would be abnormal and a sign that this is not a user looking to book a flight but more likely a bad actor attempting to commit fraud. They're likely abusing business logic, taking advantage of vulnerabilities or exploits.

So while some tools pose constraints, others can be repurposed and useful for vulnerability management.

Ultimately, the idea of constraints ties back to where you're going as an organization and what you know today. Will the tool set you have right now be enough down the road? For example, if you're not in the cloud right now but know that you'll be moving to the cloud over the next 12 to 18 months, it's time to think about the tool sets you're using to assess vulnerabilities, identify assets, and so on. Will they work in the new environment or do you need to expand your toolset?

Knowing where you're going, can your current toolset meet all those requirements? What tools are available in your organization that you're not using but could benefit your vulnerability management work?

6.2 ON PREMISE VS CLOUD

There are major advantages to building your risk-based vulnerability management system in the cloud, although there are exceptions. Both cloud and on-premise systems have all different kinds of security considerations, but how you build out security will differ depending on what you choose.

Obviously if you build in the cloud, you're no longer horizontally constrained. You can scale as you grow, rapidly and easily, often at a lower cost. More often than not, the cloud is where organizations want to build, and that would be our recommendation as well.

However, there are certain situations where organizations opt for an on-premise system. And the decision often boils down to security concerns.

It's important also to know what data is going into your system. It's one thing to say you have all your asset information and all your vulnerabilities and you're going to protect it whether it's on cloud or on premise. The ramifications of compliance are an entirely different story.

If you start to pull in application security vulnerabilities, for example, what's the output coming from those tools? Some of the appsec tools will actually capture everything that gets returned from the application and save it. That could be HIPAA-related data or PCI data. That's true whether the data is on prem or in the cloud, but it has different compliance implications depending on where it's stored.

From time to time, some organizations don't want their vulnerability data in the cloud. Federal agencies, for example, are a bit more allergic to storing asset and vulnerability data in the cloud. Some FedRAMP certifications enable them to use the cloud as well, although that requires two years of work in some cases, so it's a big consideration.

Generally speaking, with all these considerations in mind, it's more difficult to be on prem today. If you can build in the cloud, you should.

6.3 PROCESSING CONSIDERATIONS

When building a vulnerability management program, you not only have a big data problem, you have a messy data problem. You have all the vulnerabilities and all the assets you're tracking from multiple tools.

You need metadata about each vulnerability and about each asset, and that metadata has to be gathered from different places. It could be from threat feeds. It could be from sensors you're hooked into—your IDS, your IPS. It could be data from your asset management systems.

And the data changes at different rates. Some might update daily or weekly. Some will update hourly or near real time. The question is, as you're changing data, how often are you rescoring vulnerabilities? How often are you rescoring the assets? How often are you reprioritizing what needs to be fixed? How often do you update your SLAs, alerts, reporting, and all the things that matter?

The more often you're doing these things across all these different data sources at scale, the more compute resources you'll need.

6.3.1 Speed of Decisions and Alerts

It's best to start slow and simple and build toward a more robust RBVM program. At first, you might start with a simple daily update of vulnerability scores. Over time, you can increase the speed of rescoring to hourly, and then real time, before advancing to making forecasts about which vulnerabilities will pose the most risk. Obviously, as you mature and rescore at faster and faster intervals, it's going to become more costly to reconcile the data from all those different sources.

Everything should progress incrementally. A lot of organizations start by focusing on efficiency over coverage, meaning they want to get as much bang for their buck as quickly as possible early on. As they mature, they start to move toward coverage to reduce their unremediated risk as they build out their RBVM program and system.

Start by gathering data on your assets and vulnerabilities and one threat feed, and combine all that together to score vulnerabilities on a daily basis. Go in, pull all the data, rescore everything and reprioritize, and keep it simple.

Once that system is running smoothly, you can start to add additional data sources over time. You'll also want to advance from daily updates to hourly updates and get more and more robust, and add more and more sources of data over time and get closer to real time. The pace at which you advance depends entirely on your organization, resources, and where you're going as an organization. It also depends on your industry and risk tolerance.

For example, there might be a point at which daily rescoring is as far as you need to go. If you're a midsized enterprise in manufacturing, daily updates are probably OK. If you're in financial services, on the other hand, real-time and predictive scoring are necessary to address the threats organizations in that industry face. Where you end up will ultimately be a balance between your tolerance for risk versus what you're willing to invest in the cost of processing all that data.

Some organizations, like the ones we serve at Kenna Security, take advantage of economies of scale. Kenna has built a platform that many organizations use, which decreases the overall cost they would spend on processing if they were building the system themselves.

But even Kenna is a startup of 200 people. We built a system intended for use by hundreds of organizations, but we don't necessarily need to use the system ourselves in real time. We have such a small number of vulnerabilities to worry about that we might be able to just fix them all and not worry too much about the threat landscape. But many other organizations with many more vulnerabilities don't have that luxury. So processing considerations are ultimately a trade-off on processing vs cost and how much data you need and how often you need to update scoring and prioritization. All of this will evolve over time as your organization matures.

There are big data problems and bigger data problems. Obviously you would have a lot more data running a company like Facebook at scale than you would ever have in an RBVM solution. But the messy data problem associated with vulnerability management is very real. There aren't a lot of standards and there are a lot of different data sources. Sometimes they describe the same thing in multiple ways. They certainly all have their own priorities and scoring severity systems.

The aggregation and correlation involved in building out a system for scale is more complex than most people realize going in. At Kenna Security, we spent the better part of the first two years doing nothing but that. The complexity of the data model is best understood by looking at how it's structured.

The data model and your level of maturity feed into the speed of alerts and ultimately your decision-making. And once you build up to real-time or predictive alerts, you'll also have to decide as an organization what the threshold is for prioritizing remediation and how to best prioritize what needs fixing with the limited resources you have to fix them.

6.3.2 SOC Volume

Rescoring, and the speed at which you rescore, has cascading effects. New data comes in that changes scoring, issues new alerts, results in new reporting, and shuffles priorities in terms of the order in which you should fix vulnerabilities, which in turn changes SLAs.

You need to think about all of that in advance and how much you're willing to shift those priorities under the feet of your remediators. When you get to real-time alerts and decision-making, that's the sign of a mature RBVM program. But you also get to a point where you're going to make the lives of your remediators very difficult. If you tell them one vulnerability is the priority right now and then an hour later tell them that's no longer a priority and to focus on a different vulnerability, you're going to have a very upset remediation team.

You have to know what your organization's tolerance is and how much you can shift priorities beneath their feet. At some point, the patching and the fixing need to get done and every organization is working with a limited capacity of resources. And to some degree, you need to let them get to work and fix things before redirecting them to other vulnerabilities.

The concept of risk-based SLAs can help. They take into account how important the asset is, how risky the vulnerability on an asset is, and then assign a certain number of days to remediate that vulnerability.

Some vulnerabilities will be superhot and must be fixed immediately. Others you should probably never fix because they pose a very low risk and there are better uses of the time you'd take to fix them. And there's a spectrum of urgency in the middle. There will be vulnerabilities that pop up in need of urgent remediation and supersede other fixes already on the list, but knowing when to reprioritize is a decision every organization has to make.

For example, if you have a vulnerability to fix and under your SLA it should be remediated in 30 days, and the risk score for that vulnerability changes the SLA to 25 days, you probably don't have to update the due date for that vulnerability. But if the risk score changes so much that the threshold now changes to five days, that's something to reprioritize. Beyond the processing considerations of speed, there is also the speed and capacity of the resources on your team to consider and make decisions about before you put the system into action.

6.4 DATABASE ARCHITECTURE

The database architecture for your RBVM system should focus not just on the size of the database and sharding, but on user experience, factoring in things like caching and indexing and searchability.

The size and sharding of the database we've covered already, but are obviously critical to your ability to collect and process the data and knowing how much you need.

The speed to the end user is just as important. Caching and indexing matter. How searchable does the data need to be and how flexible does that searching need to be? Do you need some sort of interim layer between the database and the user? The answer is most likely yes.

6.4.1 Assets Change Faster Than Decisions

As we discussed in Chapter 3, it's impossible for most organizations to achieve 100% coverage of their assets. Every scan for assets is at best a snapshot in time and is likely different an hour later. That's OK. As we noted before, don't let the perfect be the enemy of the good.

But it's important to keep in mind that even as your organization matures its RBVM program and scales its system, bringing in more data and processing scores in real-time or issuing forecasts, the speed of your decisions will never outpace the changes in your assets.

This is why it's so critical to mature your system so it's as predictive in its scoring and prioritization as possible. That will prevent you

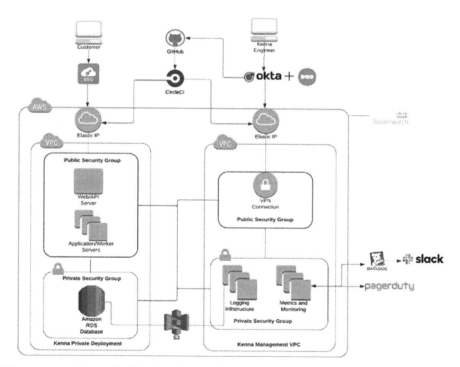

Figure 6.1 Example of a database architecture. (© 2020 Kenna Security/Cyentia Institute. Reprinted with permission.)

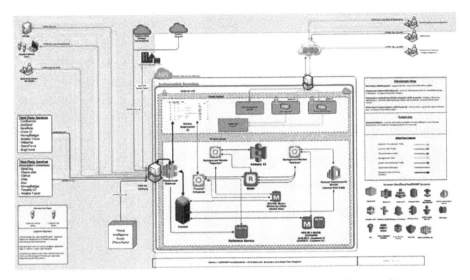

Figure 6.2 Example of a data architecture. (© 2020 Kenna Security/Cyentia Institute. Reprinted with permission.)

from wasting time remediating a vulnerability in a cloud instance that has already been turned down.

6.4.2 Real-Time Risk Measurement

We've talked a bit about the benefit of being near real time and knowing when you need to quickly react to and fix a high-priority, critical vulnerability, prioritizing it above others. But you can also go a step further.

Again, measuring risk should be an incremental process. Take baby steps. Get your vulnerabilities and assets together and add some threat data. Take a basic approach to prioritizing and see what has changed and update the priorities on a daily basis. From there, you can start to add more data sources and increase the speed of the updates.

Since you never know when the decisions about what to patch will be made, it's important to keep as fresh an inventory as possible. The only issue in traditional vulnerability management is the trade-off between real-time data and cost. In application security use cases, real-time pipelines are much more important. In fast-deploy situations, the approach needs to focus on high-risk vulnerabilities first, in real time, then run real-time checks at deploy to make sure nothing going live is crossing the business's risk thresholds.

Once you get to near-real time and you're very thorough in the number of sources you have, there's one more step to take.

6.4.2.1 Vulnerability Forecasts

The peak of maturity is to become even more proactive in prioritization. Instead of simply reacting to everything happening in the wild as quickly as possible, you can start predicting what's going to happen next.

That way you can fix vulnerabilities before exploits actually appear. Whether it be attacker trends, vulnerability forecasts, or something similar, you can start to create a predictive model. Based on all the attributes of a vulnerability, that model will be able to determine the probability of an exploit appearing for that vulnerability in the next six months, for example.

When an exploit comes out, your model might determine that you'll start to see attacks in the wild within two weeks. With the predictive model, you have a head start and can get ahead of those exploits by patching and remediating the vulnerability before any attacker has a chance to take advantage and leaving your organization in a much better place with respect to risk.

For real-time risk measurement, you need to collect a vast number of data points from disparate sources—different classes of threats, different classes of vulnerabilities—in order to get the data set that you need to make those predictions.

You also have to think about the data science pipeline to all this and the models you need in place to make these predictions. This is what we discussed in detail in Chapter 2.

At this stage of maturity, you're a different class of organization. You will have data scientists dedicated to security and to vulnerability management. Truthfully, not a lot of organizations have reached this stage of maturity today, apart from some of the big banks and major tech companies.

6.4.2.2 Batch Where Acceptable

Also be mindful of timing. As new data flows into your database, do you need to update everything at once or can you save processing power by batching the rescoring? For example, as you're adding a new source with new data, you might run a rescore job once at night or twice a day, batching it all together and pushing out a new score.

Depending on your maturity and your risk tolerance, determine your requirements. How close to real time do you really need to be?

Obviously the ideal state is always real time, but also all the compute resources we discussed earlier, including the associated costs, all go up as you get closer and closer to real time. So, especially early on in your journey, consider batching where appropriate.

6.5 SEARCH CAPABILITIES

Search capabilities make your RBVM system's user interface more flexible, so users can find what they're looking for, but also faster, because all the data is cached and indexed in a way that makes it snappier. Search also starts to introduce self-service reporting, allowing your organization to do more with the same number of employees. There are a variety of options for search, all with different licensing options and support.

When looking at search options, consider who will be using the search, the licensing options available, and how well it can scale with your organization. Some good options include Elasticsearch, Solr, and anything using Lucene syntax.

Kenna Security chose Elasticsearch based on the complexity of our data model and the service capabilities. Elasticsearch also includes natural sharding, which makes it much faster and also allows us to break it apart for different clients. If you're building your own RBVM system, you might care a bit less about that because you may have fewer worries about sharing data.

6.5.1 Who Is Searching?

To determine what kind of search and reporting capabilities you need, you also need to know who will be doing the searching and using the platform. At a less mature organization, it might just be the security team.

As your organization matures, access and self-service might expand to the IT team and operations team, and others, in which case it can be helpful to create a layer between the data and the users to offer both a user-friendly interface as well as offer each team only the information it needs.

As we'll discuss, there are a number of different use cases that should be considered.

6.5.1.1 Risk Hunting vs Threat Hunting

Threat hunting can be a broad term in cybersecurity, but generally refers to the security team seeking out vulnerabilities before they've been alerted to an issue. While SIEM and IDS and IPS systems, among many other tools, offer alerts when anomalies are detected, threat hunting is a more proactive practice, searching for vulnerabilities before they're exploited, using what's known about attacker tactics, techniques, and processes to sniff out potential exposure.

Threat hunters search for assets and look into whether any vulnerabilities or threats exist for them. The SOC and threat hunters need the ability to search your assembled data, so you need to design an interface data model and search capabilities with all of that in mind.

Risk hunting, on the other hand, is more from the vulnerability management side, which is likely where you're going to start from when you're building out your RBVM system. In this case, the team is approaching search from a different direction. Instead of searching assets and vulnerabilities for threats, the searcher starts by identifying a vulnerability and searching for the risks it might pose to your organization.

They will want to know where that vulnerability is, if anywhere, in your systems, what assets it lives on, and whether those assets are reachable by an attacker, among other questions. This requires different search functionality.

6.5.1.2 Reporting as a Service

As an organization matures, it will tend to move towards a self-service system. Organizations often start with the security team or the vulnerability management team highly involved in the RBVM system in every respect. Especially because it's new, the team knows how best to use it and wants to establish best practices early on.

In an ideal state, however, you would get to a point where operations and development and other departments of your organization don't need the security teams to understand their marching orders.

With the right search functionality, security will be able to dedicate its resources to other areas while operations and IT can use search and the user interface to understand what patching and remediation they need to prioritize. They add the items to their list and move on. They can perform their own self-reporting. As a security team, you could subscribe different teams to different reports that matter most to them. The operations team might care about some metrics and developers might care about others. You can customize reports to those needs and automatically push them to each team without burdening the security team.

6.6 ROLE-BASED ACCESS CONTROLS

A self-service system is a great goal to have, and it's usually divided up by teams based on what they're responsible for in terms of remediation. You might have a systems administrator group responsible for all the Windows systems in the organization. You might have a group responsible for the Linux systems in your data center. A team responsible for the cloud, and so on. But basically teams will be divided based on the particular set of assets that they're responsible for.

For example, if you have an EC2 instance in AWS, you might have one team responsible for updating the operating system, and another team that's responsible for updating Java or the web servers or the app servers, and so on.

Think about access controls in terms of what those teams are responsible for and ensure that your data model is capable of subdividing in that way. Build your role-based access control around those divisions. That way, if the systems administrator for the cloud environment logs in, they're going to see all of the instances that they're responsible for, all of the operating system vulnerabilities they're responsible for, and nothing more.

When the app server team logs on they can see all of their logic, all of the Apache web servers, and so forth, and nothing more. But you need to be able to make sure that the data model is capable of doing that. The role of the back-end system you build on top of your data model should be capable of filtering out that view to just the relevant users.

As we've seen throughout this chapter, building and scaling an RBVM system is a process that can take years to reach full maturity. There are a number of elements to consider before you get started, and with your organization's overall IT roadmap in mind, you can implement the compute, database architecture, and search capabilities that get you off the ground and give you room to grow. This incremental approach to maturity is just as important in bringing your integral processes and teams into alignment, the subject of the next chapter.

7

ALIGNING INTERNAL PROCESS AND TEAMS

Every organization begins the journey to RBVM from a different state. Some organizations, like banks, insurance companies, or healthcare organizations, are already attuned to risk and have a smaller leap to make. Others are new to the concept and teams will need some convincing to shift from well-trod processes and procedures.

Every organization progresses through several phases of maturity. Some are starting fresh, gaining visibility into the high-risk vulnerabilities that have been lying dormant in their environment. Others might have fixed the worst of their vulnerabilities but still need to operationalize their RBVM program with metrics and SLAs that help them maintain an acceptable level of risk.

No matter where you're starting, you need to make sure your teams are on board and aligned around a common goal and establish processes that convert intelligence and decision support into action. Done right, you'll have a more efficient team that's actually moving the needle on reducing your organization's risk.

Understanding this maturity process and what you're actually trying to achieve is an important part of a successful RBVM program. What looks impossible when you first discover the millions of vulnerabilities in your environment becomes manageable when prioritized based on risk. Through the maturity process, you will come to better understand attacker behavior using data science, refocus your efforts around risk by quantifying goals and operationalizing RBVM, and achieve a high-performance steady state that keeps your risk low and measures your effectiveness and progress.

In this chapter, we're going to explore the application of RBVM through people and processes that transform your vulnerability management and the relationship between IT and security to attain superior outcomes with less effort. We'll look at the risk-based goals and metrics you'll need to track and how to convince IT and DevOps teams, as well as your organization's leadership, that this is the best path forward. From there, we'll look at the different stages of maturity and the processes and team alignment required for success at each level. Done right, you'll follow a journey that lowers institutional risk while achieving savings in cost and time.

7.1 THE SHIFT TO A RISK-BASED APPROACH

While organizations arrive at RBVM from a wide variety of challenges or use cases, they typically have some common goals in mind. At this early stage, it's important to clearly define those goals so they're easier for the team to understand and act on.

7.1.1 Common Goals and Key Risk Measurements

Every organization has unique needs that will shape their goals, but those goals tend to fall in one or more broad categories:

1. *Keep up with the growth in vulnerabilities.* The average enterprise has millions of vulnerabilities, and the volume and velocity of them continues to skyrocket. It's impossible even for well-resourced security and IT teams to fix them all, but they need to address the vulnerabilities that pose the greatest risk.
2. *Automate manual VM processes—particularly prioritization.* Security's traditional approach to identifying and prioritizing vulnerabilities is largely manual and still leaves IT and DevOps with far too many vulnerabilities to remediate. The decision engine we built in previous chapters will automate prioritization of vulnerabilities and make everyone's job more manageable.

3. *Improve efficiencies.* Security teams can spend a lot of time investigating and monitoring vulnerabilities, and then it's up to IT and DevOps to remediate the most important ones. The new efficiencies of RBVM can help organizations make better use of existing resources and budget.
4. *Reduce friction between teams.* When security hands IT and DevOps endless fix lists with no explanation or understanding of priorities, they're not exactly thrilled. At a certain stage of maturity, you'll be able to hand the reins to IT and DevOps to find and fix the most critical vulnerabilities.
5. *Streamline compliance.* Virtually every large enterprise is required to meet various data security requirements, and those that handle sensitive customer data or payment information face additional scrutiny. Easing this burden is almost always a goal.
6. *Improve communication and reporting.* Board members and nontech C-level executives generally don't speak vulnerability management (VM), and old-school metrics are largely meaningless to them. Chief information security officers (CISOs) and other security executives are always looking for metrics that quickly demonstrate progress and success.

To reach these milestones, you need the right metrics and key performance indicators (KPIs) to measure, demonstrate, and communicate your success.

As we've discussed in previous chapters, these metrics should focus not on vague goals like how many vulnerabilities you can remediate in a certain period of time, but reflect the reduction in risk the team's actions have achieved.

As we've discussed in Chapter 5, those metrics can include:

- *Coverage*, which measures the completeness of vulnerability remediation, or the percentage of exploited or high-risk vulnerabilities that have been fixed;
- *Efficiency*, which tracks the precision of remediation, such as the percentage of all remediated vulnerabilities that are actually high risk;

- *Velocity*, which measures the speed and progress of remediation;
- *Capacity*, which determines the number of vulnerabilities that can be remediated in a given timeframe and calculates the net gain or loss;

Communicating these metrics to executives, however, is not always smooth. It can help to have a simpler metric, an overall assessment of your organization's risk.

Risk scores and risk meters are two ways to do that. By harnessing the variables more indicative of the future weaponization of a vulnerability, we can quantify the risk any vulnerability poses to your organization.

Risk scores, based on a scale of 1 to 100, with 100 representing the highest risk, are assigned to each vulnerability. That number represents the relative risk the vulnerability poses to your organization based on a number of factors, including the prevalence of the asset on which it exists, the likelihood it will be exploited by hackers or malware, and your organization's own tolerance for risk. The risk score helps security teams prioritize and manage that vulnerability, and it gives them an evidence-based way to align IT and DevOps around the same priority.

A risk meter allows you to instantly view your progress in reducing risk by department, asset group, or other category. At early stages of maturity, risk meters can motivate teams into a friendly competition to see which group can drive down its risk meter furthest. Risk meters are particularly helpful in communicating progress to nontechnical audiences, and help ensure adoption of your new RBVM tools and the larger program, which in turn leads to security achieving greater ROI from its investment.

In most cases, changes in the risk meter or risk scores are easier to communicate than demonstrating that you've prevented data breaches, denial of service attacks, ransomware infiltrations, and more. They align teams and can bring executives on board with the RBVM approach.

7.1.2 Case Study: More Granular Risk Scores for Better Prioritization

A major consulting firm took its risk scores a step further to better guide its teams. A global organization, it had a strong and clear sense of the risk levels it was comfortable with and effectively distributed that methodology across all its member firms. The security team consistently measured progress and divvied up what each business unit should focus on from a remediation perspective.

This company broke down the high-risk vulnerabilities into several categories for a more granular sense of what is high risk to better prioritize vulnerability remediation. The category of high-risk vulnerabilities, for example, is subdivided into critical, high, medium, low. As soon as the critical high-risk vulnerabilities are taken care of, the team moves to the next tranche. It's been incredibly successful simply because they've been able to prioritize the highest-risk vulnerabilities and develop a classification system to ensure teams are focusing their time, attention, and resources on what matters most.

By looking at its many vulnerabilities through the lens of risk, this company has reached a point where the majority of its vulnerabilities are in the lower medium to low category. Critical vulnerabilities have been driven down to zero. When new ones pop up, they're raised to the top of the priority list and quickly remediated in line with their SLAs (more on SLAs later in this chapter).

7.1.2.1 The Importance of Culture in Adopting RBVM

Certain organizations have a head start on a risk-based methodology. Banks, insurance companies, even healthcare to a degree, have risk in their wheelhouse and the challenge is simply how to translate that into vulnerability management. If you have a strong risk management path in your organization, your teams likely understand those routines well and it's fairly easy to add RBVM to the mix because there's a fundamental understanding there.

For organizations that already have a good process for risk management, they just need a way to talk about RBVM with real quantified risk that they can embed in their cycle and go chase. It works well because you don't have to convince anyone of the soundness of the approach.

Other organizations need to have conversations with their team and level with them that the process they're currently using hasn't been the most effective. In this case, you have to convince them that you can cut the number of vulnerabilities that need patching by an order of magnitude, if not two. It's going to save everyone time, and everyone can access the data to see exactly what needs patching.

In some cases you then have to go talk to the auditors and convince others. You have to sell the value of savings in time, energy, and avoided downstream risk on security. It's not a small change; it's a different approach to risk that changes the way you operate.

Without the right relationships, there are going to be battles. In one organization, Java ended up in a limbo state due to poor relationships. It was a risky part of the organization's environment, but because of internal battles and politics, it was never fixed.

For any organization, the speed of adoption is determined by the culture inside the company—what the relationships look like. Is security viewed as a strategic function? Do you have a strong mechanism for dealing with risk in any form?

We're often asked about the best team structure for RBVM, but team structure matters less than how teams work together. The right relationships empower the process change. Any team structure could conceivably work. It's not a determining factor.

What actually determines success or failure is how the teams work together and whether the relationships are in place to make the team care about the change and take action based on it.

7.2 DRIVING DOWN RISK

It's time to say goodbye to spreadsheets and reducing vulnerability counts. You're probably going to discover that you've been taking actions that had no meaningful impact on risk, and potentially created risk by focusing on the wrong vulnerabilities to remediate. That's normal. You're taking the right steps to efficiently reduce risk by focusing on what matters.

When you first get started in RBVM, there are often several surprises: the number of assets in your environment, the number of vulnerabilities in your environment, and most important, the level of risk in your environment. For any organization just getting off the ground, your risk is going to be much higher than whatever you've determined to be an acceptable level. So begins the first phase of RBVM maturity: driving down risk.

Recall the survival analysis charts in Chapter 5. The organizations included in that chart on average remediated a quarter of vulnerabilities in about four weeks, half in the first 100 days, and more than a quarter stretching out to over a year. When remediating high-risk vulnerabilities, you'll see a similar initial velocity to your risk reduction. Once you've eliminated the most critical vulnerabilities, the velocity will start to taper off as you approach your acceptable level of risk.

This is the dual challenge of risk scores to keep in mind when you're at the earliest stage of RBVM maturity: Initial risk scores will be quite high, and your focus on driving down risk will eventually reach a point of diminishing returns. All of this is normal.

When you first measure the risk in your organization, you'll find there's a lot of work to be done. That makes sense—that's why you're pursuing an RBVM program in the first place. But the high score can be demoralizing for teams and concerning for executives.

Once you've driven down risk to an acceptable level, the progress you've made and the reduction in the risk score you've achieved will plateau, eliminating the dopamine hit of watching organizational risk fall with every remediation.

At a certain point of maturity, you'll reach a steady state where it's harder to measure your performance based on the risk score. At that stage, it's more important to measure the response to risk—are you staying at your goal score and how fast do you return to that goal score after you see a spike in risk?

We'll get to all that later in this chapter. The important thing to remember for now is that your goals need to shift.

7.2.1 Aligning Teams with Your Goals

If you're just getting started with RBVM, your security and IT teams have likely been working toward different objectives. That's why corralling everyone to shared objectives, such as a risk score, is so important. It ensures everyone understands their role in the process, what they're tasked with, and why.

Every time IT steps in to help security, the team takes on risk because they're making changes to the environment. Working from shared objectives will reduce the amount of change required and the amount of risk IT takes on. It doesn't just reduce risk or the amount of work IT has to do, it also saves time that would typically be spent deciding what to do. It amplifies every action, since every action meaningfully reduces your organization's risk and helps both teams achieve their objectives with less time, energy, and effort.

Without collaboration, it's harder to implement the methodology and the mindset. You need to break down the barriers between the individual teams responsible for doing the work. Part of that is education.

The first step is familiarizing the team with the metrics and goals you're pursuing. Explain the risk score and the methodology behind it, as well as the baseline for where you are right now.

Walk through their questions and address their concerns and any roadblocks to shifting a new process flow. This is where the science and methodology play a major role in bringing the team on board. Once they believe in the new process, repetition and practice are needed for the team to fully adapt to it.

7.2.2 The Importance of Executive Buy-In

A lot of teams struggle with this early stage of RBVM. The most important element in aligning the team with this approach is to win over leadership. If your leadership isn't on board with accepting this method, the teams aren't going to do it. It's key to get your metrics in front of leadership to paint a picture of risk posture for them.

It's important also to highlight the risk of inaction. Every organization has critical assets and good arguments for not patching them, for fear that the patching process will create a much bigger problem with an application that's critical to the business. But if you can show the cost of not fixing it, of living with the risk that a particular vulnerability poses, you can earn leadership buy-in.

A change in leadership can be harmful or helpful to your cause depending on the attitudes the new leader brings to the organization. Sometimes you spend time and effort winning over a leader only to see them leave and be replaced by someone skeptical of the risk-based methodology. You might hear statements like "we have our own threat intelligence," or "we're smarter," which can be frustrating. You'll have to start the process from the beginning, winning over the new executive with data.

On the other hand, a change in leadership can also accelerate the adoption of RBVM. Some of the best times to pursue RBVM as a new initiative are at moments of change. When new leaders step into place with a fresh vision and an eye for improvements, RBVM can quickly catch on. When organizations are in a transitional state, introducing new processes might require less effort.

Focus your efforts on winning over leadership. Without their support and directive, it's going to be much more difficult to get the entire team on board.

7.2.3 Reporting New Metrics

Which metrics will help make the case for RBVM to leadership, and how often should you present them?

It's good to deliver some metrics on a weekly basis, especially when reporting to the chief information officer (CIO) or other C-suite management. One is a risk score associated with each area of focus. If the score isn't aligned with the agreed-upon risk tolerance, the reporting should also include an explanation of how many new vulnerabilities came up that week and what the plan is to fix them.

Other metrics to present include the risk score for direct reports and the risk score over time for the company as a whole. Are the teams fixing the right vulnerabilities? Are you moving away from vulnerability counts and looking more at the risks?

If you have an existing process in place, you can make the transition more seamless by sprinkling in risk-based metrics slowly. For example, one organization legally had to remediate vulnerabilities based on Qualys scores. Over time, the team was able to introduce risk-based metrics on top of the current process, building in RBVM to add some risk-based color.

As soon as everyone's aligned with the RBVM methodology and driving toward the same score, you're on your way. But you have to get management buy-in and these metrics are key. Trying to keep everyone on the same process and changing the platform, the data, and the algorithm slowly over time can work the team toward transitioning to RBVM.

7.2.4 Gamification

Once you've established that risk scores are worthwhile and converted the team to using them, you might consider setting goals and targets tied to compensation or publicly calling out the teams that haven't been keeping risk on track. Gamification too can bring teams on board and focus them on new metrics for success. Some organizations use this level of competition and peer pressure to keep risk scores in the right range.

By tying success to variable compensation, you incentivize teams to take a specific action. They either have to conform or don't get the extra pay. That has worked well for some organizations trying to bind remediation teams to certain success criteria.

The ultimate goal with any incentive program is to get teams to coordinate and work together to manage risk. It releases some of the tension that teams typically have across departments and levels the playing field by clarifying what everyone is marching toward.

However, it works best only once you have the processes in place to empower teams to understand which vulnerabilities to tackle and why. In some cases, when you're just trying to stand up your vulnerability management program, incentives and gamification may not work because the lack of defined processes and workflows can cause frustration.

Gamification is most productive in this early stage of RBVM after you establish a score and are working through it. At a certain point, it becomes counterproductive. Risk reduction can be addicting, and taken too far, the return on your effort starts to plateau. But a renewed focus on performance can make sure you're remediating vulnerabilities as close to optimally as possible.

Which is why you need to start shifting teams from focusing on risk scores to thinking about performance. Once you reach full RBVM maturity and achieve a steady state, risk reduction is not going to happen. At that stage, you need to ensure everyone is doing a good job of remediation, which is where remediation scoring comes into play.

Everyone has to fight through the first phase to get risk down to a reasonable level. This is where leadership buy-in and team alignment are so important. But where do you go from there?

How do you think about performance in response to risk in a steady-state scenario? Are you taking the right steps fast enough? Do you understand your risk tolerance and how that is reflected in your SLAs?

7.3 SLA ADHERENCE

Once you drive down risk, you'll need a new goal to demonstrate progress and continue to show leadership how far along you are. Many organizations don't even know what a "good" level of risk looks like or how to operationalize RBVM, let alone whether there's a phase beyond that.

Most companies will get to a place where they have a risk tolerance and SLAs and that's how they operationalize the process, gauging performance against those metrics to continue to make progress over time. As the team matures, those requirements might tighten further.

The beauty of security is we're never done. Once you put in place a process that's driven by risk and operating at your accepted level of risk, you can achieve a steady state where the most efficient organizations are remediating roughly 25% of vulnerabilities that come into their environment every month, including the ones that pose the greatest risk to their organization.

Operationalizing the RBVM approach and the next step in maturity begins with SLAs.

7.3.1 High-Risk vs Low-Risk Vulnerabilities

The ability to distinguish between low-risk and high-risk vulnerabilities reduces your field from 100% to 2% to 5%, turning an intractable problem into one you can actually accomplish. Moving fast to remediate just that 2% to 5% of most critical vulnerabilities also makes you more effective at reaching the level of risk your organization is comfortable with.

Moving to an RBVM program involves shifting your focus as an organization—you need to change the way you think about and remediate vulnerabilities. But how you measure success can be an overlooked aspect of shifting your teams' focus to high-risk vulnerabilities.

Many companies report vulnerability counts and the percentage of vulnerabilities at different risk level scores in their environment. Shift your reporting to the risk scores, as we discussed in the section above. Teams can patch vulnerabilities all day long, all week long, all year long, without ever making any dent in the risk your organization faces.

A global financial company was in exactly that position. It had been prioritizing its remediation and reporting based on Qualys severity scores. Unfortunately, the vulnerabilities up for remediation included a lot of false fours and fives, and the team couldn't keep up. The reporting reflected this losing battle.

At one point, the team reported to the company's board of directors that they had remediated 1,500 vulnerabilities, but another 3,000 had piled on in the same time. It was difficult to demonstrate progress and the effect their work was having. Even more problematic for the team, the Qualys scores were written into their remediation policy. That was the bar they had to reach, and it's very difficult to change that kind of policy once it's written.

The company switched to a risk-based vulnerability management program, applying threat logic to better target its remediation efforts. Now it has a better finger on the pulse of which fours and fives pose the most risk, enabling the team to tackle the ones that actually matter while still working within the company's written policy.

The team has been getting kudos internally for how mature the program is becoming and the teams are more willing to shift their methodology because they have actual data behind each action they take—it's not just another false four or false five. Over time, the company plans to rewrite the policy, but for now the teams are starting to socialize the risk score and the reduction in risk that their actions are having. They're showing the value in the new approach and once its success is demonstrated and the teams have confidence in it, the company can formalize them.

Focusing on vulnerability counts will never have a consistent, measurable impact on lowering risk. Instead, focus on the worst, riskiest vulnerabilities driven by a risk-centric approach and supported by SLAs that set shorter remediation timeframes for the most critical vulnerabilities. Reporting helps to clarify what's important and also convey to executives up the chain of command that the efforts are successful.

7.3.2 When to Implement or Revise SLAs

Most organizations start down the path to RBVM with SLAs already in place. Unfortunately, teams generally aren't hitting them, or at best, they hit some and miss others. This is due at least in part to the goals they're being asked to achieve. As we just discussed, reducing vulnerability count is always a losing battle since no matter how many vulnerabilities you resolve, no matter how fast you do so, there will always be more. It's next to impossible to consistently drive the vulnerability numbers down and make progress.

7.3 SLA ADHERENCE

By shifting to a risk score and risk score reduction, the target number becomes nonzero. In the initial phase of driving down risk discussed in the previous section, you shouldn't worry about SLAs because there are higher-risk vulnerabilities that have been sitting in your environment for a long time and may be well exploited and well weaponized. You have to go after them at scale.

But once you've driven risk down to a range where you're starting to see a diminishing return, risk reduction ceases to be useful. Once you've gotten leaders and the team on board with the new process, the process is underway, and you've tackled the most critical vulnerabilities; then you should put SLAs in place to continue your success by tracking and measuring progress via SLA adherence. Reducing your risk further is counterproductive.

Based on your organization's acceptable level of risk, set different tiers of escalation based on the risk score of different vulnerabilities. As we saw above, some companies set very granular levels of vulnerability risk, breaking each tier into subtiers to better classify priorities. However you've chosen to rank vulnerabilities by risk score, set SLAs for the number of days each tier should be allowed to remain open based on that level of risk.

Keep in mind that the risk score isn't as important as your overall risk posture. As long as you're responding to risk appropriately within SLA timeframes, the risk score of any one vulnerability is actually less important. What is important is that you're meeting your SLAs and staying within your acceptable level of risk.

The SLAs you set will look different from the SLAs you had before starting down the RBVM journey since your new SLAs will be oriented to high-risk vulnerabilities. Low-risk vulnerabilities will have a long remediation time—the SLA might be to fix low-risk vulnerabilities in a year. In some cases, you may never need discretionary effort to meet an SLA for low-risk vulnerabilities, since service packs will pick them up. Focus on the high-risk vulnerabilities and don't go out of your way to accelerate your effort for the low-risk ones.

The trick with getting up to speed on SLAs is that in many organizations, there are vulnerabilities from long ago, possibly 20 years or more, that have been sitting in your environment. The teams are way over the SLA on those right out of the gate, which is a deterrent to taking action for fear of C-level executives questioning why the company has vulnerabilities that are 20 years old.

This is why you should introduce SLAs after you've remediated all your higher risk vulnerabilities. If you start behind the eight ball, you're always going to be behind. If you're working at a level where you can keep up and can track that teams are staying on top of risky vulnerabilities, it's easier to sell, track, and measure the new approach.

Table SLAs until you have the basics covered. Understand where you're starting and see if the SLAs will help or hurt you and move toward them over time as you remediate the worst of your vulnerabilities. You can try to put in place risk-based SLAs that make you faster than your peers but you're never going to be faster than your attackers, so if you have vulnerabilities that are 10 years old, you have to clean them up first.

7.3.3 What to Include in Your SLA

Getting a risk-based score into an SLA policy is key. Many teams don't put this in their policy, but it's important for auditing and compliance. When auditors come on site, they're going to ask for the security policy and whether following SLAs is part of the policy. If you don't have a policy, they'll tell you what they think you should be fixing. It's important that when you reach the right stage, you put your RBVM approach into policy, document it, and make sure you've made clear what is critical and what is not.

Because SLAs are an agreement between security and the IT organization executing the remediation, the clock starts ticking as soon as the ticket is cut. Before then, the IT organization is largely unaware of the issue. In smaller organizations where remediation and security might be the same person, SLAs might start as soon as discovery. That said, even in the most sophisticated organizations, there's usually a day or two of organizational delay before the SLA clock starts. As long as this time is minimized for the small percentage of vulnerabilities that are high risk, delays of a day on top of an SLA of 500 days aren't so bad.

You should be able to clearly show SLAs for the internal network vs external. An externally facing asset will have a shorter SLA than an internal one.

You can also start to break down your CMDB metadata around different compliance requirements. For example, you could have certain timelines for fixing SOX-applicable servers, or an internal PCI server, with a specified number of days to remediate any vulnerabilities.

You have to make sure the SLAs are documented policy and provided to your users, because there will be internal teams auditing the effectiveness, external assessors coming in, and even clients that audit and look at your policies to make sure you're following those metrics.

Some organizations will require a complex SLA policy that includes regulation categories like PCI and SOX, as well as other categories, and a matrix. But it doesn't have to be that complex.

What's important to have in your SLA is external vs internal goals, because those will differ in the number of days you have to remediate vulnerabilities on either side. You should also break it up into different levels of criticality. For example, a risk score of 90 to 100 might be critical, 67 to 89 is high, 34 to 66 is medium, and anything 33 or below is low. For the low-priority vulnerabilities, you don't even necessarily need an SLA on those.

Some organizations don't use SLAs but have different processes in place. They might use risk scores and set requirements to stay below a certain threshold of risk. If they rise above that line, the team has to explain why, and create a plan and timeline for fixing it. For example, they might report that a vulnerability came in overnight, it's actively being weaponized today, and here's the plan to fix it this week. They have some alerting in place for when a risk meter goes over a specific tolerance. At that point, they send users a report and notify them that they're over.

SLAs are important especially for the critical and high-risk vulnerabilities because they keep the team on track for remediation and hold them accountable. You might be fine today in terms of risk, but tomorrow could bring a critical new vulnerability. Without an SLA, the team would keep getting dinged for being over the risk tolerance until the vulnerability is fixed. With an SLA, you can hold the team accountable to a set period of time to fix that vulnerability and leave them to fix it within the designated time frame.

7.4 SHIFTING FROM SECURITY-CENTRIC TO IT SELF-SERVICE

In most organizations, IT remediation teams get marching orders from security teams. They're waiting for instructions, the sanitized list of vulnerabilities to focus on. The next stage of maturity is moving toward a self-service model where IT can see and understand why vulnerabilities are scored the way they are and why they pose more risk than others. This way, you can empower the users who are actually responsible for mitigation to be the decision-makers. This model cuts down on time, gives IT more keys to the kingdom, and empowers them to avoid the kind of roadblocks that crop up between IT and security.

The goal of shifting to self-service can be difficult to achieve. Many medium to large enterprises, with their vast scale and unique requirements, expect and deserve more hands-on attention. Another reason is that a lot of pieces have to fall into place for self-service to be successful.

Some organizations have security and IT ops under one umbrella, which makes it easier to sell the shift in approach and methodology, but often there are discrepancies between IT and security. The challenge is overcoming internal strife and getting both teams on the same page, with one tool, methodology, and process that aligns marching orders so that everyone is focused on the same goal—to reduce risk.

Vulnerability management teams tend to be small. Even in large banks and healthcare companies where teams are well staffed, you might have four people. Those four people manage all the vulnerabilities in a major enterprise. It's a massive undertaking. Having other team members be able to log in and take care of work is crucial because otherwise there aren't enough people to get the work done.

7.4.1 How to Approach Change Management

Change management is key here. The IT team is used to looking at information from a scanner, but when you introduce a risk-based decision engine, they have to be trained to use it. How successful this is depends, again, on how much the leadership has bought into the approach and endorses it.

Rolling out a new system and a new process will be stressful but ultimately rewarding. Teams will complain because they have their entrenched way of working and when you introduce a new system, they don't want to learn it. They don't care about security as much as the security team does. They're just applying patches and doing their job. But ultimately they—and you—will be happier and more successful with self-service.

One of the most challenging steps in the training process is getting teams used to risk scoring. If they're changing from using an intelligence-based methodology, it brings to light some risky vulnerabilities that they don't like to patch. The regular Patch Tuesday fixes go out automatically and that's pretty easy for most organizations. It's the application-based vulnerabilities that cause more consternation.

These vulnerabilities have been sitting on the sidelines because many organizations have been remediating based on CVSS, and there are more vulnerabilities to remediate than anyone has time for. When you switch to a risk-based approach, it shrinks the pile of high-criticality vulnerabilities, but they're often the kind that break applications. The pile is smaller but each one is tougher to patch. Getting teams used to the scoring and having them actually work on critical vulnerabilities can take time.

Ultimately what makes this approach successful is repetition. It takes time to get used to the new method and workflow, but if you stick with it, pushing teams to drive toward the risk score and remediating the most critical vulnerabilities, the team will realize the effectiveness of the approach and their own success.

When you conduct user training, record it so teams can view it multiple times. One time when we did a training for patch teams, it turned into a compliance session because they didn't fully understand and weren't comfortable with the scoring. With repetition, they're beginning to get used to it.

Repetition means meeting with teams every day, talking through what they're seeing. It takes time and patience getting them used to the data and the scores. It takes leadership pushing to fix these vulnerabilities and addressing the high-risk vulnerabilities the team didn't fix on time.

7.4.2 Enabling Distributed Decision-Making

Let's say you have 3 million vulnerabilities in your organization and the biggest chunk are Java. You're going to make a big push to reduce Java risk. But the majority of vulnerabilities probably aren't Java. The Java team shouldn't be the only one hard at work. If you delegate decision making to every team and every team understands they can look at their infrastructure and see the risk specific to them, they can then patch the highest risk vulnerabilities in their part of the business.

This approach achieves vastly more efficient and comprehensive risk reduction because everyone is actively working to reduce risk. You still get rid of your high-risk Java vulnerabilities, but you also go after other high-risk vulnerabilities. The more you can break the problem down into a smaller number of machines, get closer to the application, the more you break up the problem into manageable chunks versus having it centralized.

The key is to establish a self-service system in which the application owners or whoever owns the infrastructure will be responsible for changes. They can see the risk they have and the actions they have to take to drive the score to the appropriate levels. They make the decision, they drive action in the form of tickets or other methods, they have a shared view of what success looks like—specifically, what risk levels your organization has deemed appropriate.

Make sure you put guardrails around self-service. Simply allowing remediation owners to pick what they want to fix can lead to them avoiding the vulnerabilities that are hard to fix. Give them some leeway to choose, but make sure they're focused on the most critically important risks.

Another important piece is good metadata in your CMDB. This not only helps remediation teams do their jobs, but also allows security advisors to create different views of your environment based on audit requirements.

If you have very targeted groups of assets, you can segment by member firm, by critical asset type, and then pinpoint the crown jewels of assets that must be remediated. Being organized and knowing what you have is key to avoiding the game of "that's not my asset."

Where relevant, you should also note in your CMDB metadata whether an asset is a PCI asset or SOX application that contains sensitive data. You can bring that into your decision engine as a tag. Correlating CMDB data with your vulnerability data is another way to help the team prioritize their remediation work and its impact.

A major shipping company has a great CMDB integration, with some 800 groups that can access their RBVM software. The CMDB integration allows them to establish ownership, shifting responsibility from the security center to self-service so the vulnerability management teams are no longer a bottleneck. They use CMDB data to create specific risk meters and groups for each different area of the business. Each user has a specific role so that when they log in, they only see the relevant information for their part of the business—the risk tolerance they should be driving toward, the buckets of work they're responsible for, and when that work is due according to SLAs.

All of this depends on having a good handle on your asset inventory. A lot of times when organizations are making the switch to RBVM, their CMDB is outdated or full of garbage. If you can fix that data source and improve your metadata, you'll be light years ahead of the majority of organizations approaching the transition to RBVM.

7.4.3 Signs of Self-Service Maturity

One sign you've hit a new level of maturity is realizing you haven't used your decision engine in a while. It seems counterintuitive, but when your IT and operations teams are using it multiple times every day to find the next most critical vulnerability to patch, you as a security team might only have to check in periodically for reporting and to confirm everything's on track.

You'll know you've reached this stage when you're no longer driving the process. That's what winning looks like. In a world with a cybersecurity labor shortage and skills gaps, the ability to hand off the vast majority of vulnerability remediation and focus your time and attention on other responsibilities will help your organization become more efficient as you reduce your risk.

7.5 STEADY-STATE WORKFLOW

New vulnerabilities pop up every day. Most are harmless, but occasionally, something dangerous is released into the wild. When it happens, your risk score will jump. It's no one's fault, but if you've established SLAs and you've achieved self-service, your team will be able to step into action in line with your organization's tolerance for risk and mitigate the issue.

But there are a few additional points to keep in mind at this stage of the game.

7.5.1 The Limits of Remediation Capacity

As we learned in Chapter 5, every organization has capacity limitations. Within a given time frame, an organization can only close so many vulnerabilities. The portion they're able to close is their remediation capacity.

When we looked at the capacity of roughly 300 organizations, we found that a typical organization, regardless of asset complexity, will have the capacity to remediate about one out of every 10 vulnerabilities in its environment within a given month. That seems to hold true for firms large, small, and anywhere in between.

Smaller organizations have smaller teams, but also smaller problems. On the other hand, a large enterprise might have bigger teams able to fix more vulnerabilities more quickly, but they're working in a much more complex environment with many more assets and vulnerabilities. However, this doesn't mean that one in 10 vulnerabilities per month should be your goal.

The highest performing organizations achieve two and a half times the capacity of the typical organization—remediating just under one in four vulnerabilities.

Looking at the number of high-risk vulnerabilities in each firm's environment and the number each organization was able to remediate in a month, we saw that about a third of organizations are able to gain ground on the high-risk vulnerabilities, while a sixth keep up, and half fall behind.

The good news is that the typical firm with a capacity to remediate 10% of high-risk vulnerabilities has the capacity to remediate the 2% to 5% of vulnerabilities that pose the greatest risk. The highest performing organizations that can remediate just under 25% are even better equipped to respond as new high-risk vulnerabilities appear.

7.5.2 Media-Boosted Vulnerabilities

Every time a major data breach or hack makes the news, security teams everywhere see messages, phone calls, and emails light up with questions from executives or customers concerned about falling victim to the same vulnerability.

With an RBVM program, you'll be able to respond with the likelihood your organization will be hit and what the impact might be. That's where having metadata associated with assets allows you to clearly say what percentage of assets in your organization contains that vulnerability, which assets are in scope for PCI, which assets are in scope for SOX, or whatever the different categories or classifications are for sensitive data in your company.

Armed with this information, you'll either be able to communicate how quickly the vulnerability will be remediated under SLAs—or whether it already has been. Or instead, you'll have the data to show both executives and customers that this vulnerability, although it's gaining a lot of attention at the moment, poses little risk to your organization and will be remediated in the appropriate time frame.

7.5.3 Exception Handling

In the course of remediating vulnerabilities, you will encounter exceptions. These might include vulnerabilities for which a patch does not yet exist or perhaps the vulnerability exists on a critical system and there's a justifiable reason to avoid patching it under the typical SLA time frame.

No matter what form the exception takes, it's important to have a working definition for what constitutes an allowable exception so that when one does appear, everyone can agree it fits the bill. Defining the criteria with a narrow scope will ensure that exceptions are a rare and acceptable risk. The authority to declare a particular exception should rest with leadership.

When leadership declares an exception, rather than fixing the vulnerability, security and IT teams will need to work together to put in place alternative safeguards to protect against exploitation.

7.6 THE IMPORTANCE OF PROCESS AND TEAMS

The hard work of building a decision engine to prioritize your most critical vulnerabilities is key to reducing risk in your organization. But it's all for naught if you can't convince security, IT, and operations teams to use it and integrate it into their current processes.

Convincing leadership of the value of the RBVM approach and coaching teams through new processes are first steps. Once you've driven down risk to an acceptable level, you can shift your focus to a steady state, meeting SLAs to address new vulnerabilities as they appear.

While data science and technology are essential to fixing the riskiest vulnerabilities, people and process are just as important, and will ultimately determine how successful your RBVM program becomes.

8

REAL-WORLD EXAMPLES

8.1 A WORD FROM THE REAL WORLD[1]

When I joined HSBC, the vulnerability management program was a two-person team. They were essentially running a bunch of PowerShell scripts and using an outdated tool called IBM QRadar Vulnerability Manager (QVM) to identify vulnerabilities. They would manually review which vulnerabilities they felt were relatively critical and ticket them out to be fixed. They didn't have a good mechanism to tie those vulnerabilities to their owners. A lot of it was guesswork and random ticketing.

By the time I left HSBC, we had built the vulnerability management team to around 40 people. We'd implemented scanning from 30,000 IP addresses to over 14 million possible IPs, totaling about 1.5 million assets, and then started to implement vulnerability assessments. As the program grew, we found a lot more issues, and understanding which ones to prioritize was critical to our success. Risk-based vulnerability management gave us the prioritization engine we needed.

As we built out the program, we focused on four key areas of the end-to-end lifecycle of a vulnerability:

1. Discovery;
2. Assessment and prioritization;

---

1. This chapter contribution is by Will LaRiccia.

3. Communication;
4. Remediation.

Here's how we took HSBC from guesswork to a data-driven, prioritized approach to fixing vulnerabilities that lowered risk and has made the company a shining example of what RBVM can do.

8.1.1 Vulnerability Discovery

The first step was uncovering all the issues in our environment. That meant gathering information to achieve the coverage and depth of visibility we required.

We partnered with our networking team, our security operations team, and nearly every business stakeholder to identify our actual vulnerability footprint, as well as to understand that footprint in terms of IP address ranges and what our external estate was. All those different factors helped us identify what needed to be scanned and found.

You'll be amazed at the different types of devices you'll find on the network. We had everything from gaming consoles to connected fish tanks.

8.1.2 Vulnerability Assessment and Prioritization

Once we understood what we were scanning, we had to figure out how to prioritize what we found. We needed an asset prioritization score.

HSBC was operating in a heavily regulated industry, and we needed to know which assets were regulated. That became one of our factors for prioritization, because if a particular vulnerability was on a regulated asset and we didn't fix it in time, that could lead to fines and not being able to operate within a particular region or country.

A second factor was the network location. Any asset on the core network was a higher risk than one in what we called the demilitarized zone (DMZ)—the gap between the internet and HSBC's internal network. Being on the DMZ was lower risk because HSBC had a flat network and any device on the core network could connect to anything—including the production system processing hundreds of thousands of payments per day. That was a big issue.

The third factor was whether the asset was in a production environment, a Wi-Fi environment, a development environment, or another type of lower environment. How would we prioritize among environments? Which would we patch first?

Finally, we had to factor in the different device types beyond servers. How do we prioritize a vending machine over a heating system that has an IP address? We came up with a scorecard for those different types of devices.

Thinking through those factors and reaching agreement on how to score assets and prioritize them is never clear-cut. For example, you could argue that the security cameras in a bank are pretty critical since they help protect the money. But does that make them more or less important than an ATM, which has direct access to money?

In these kinds of conversations, it's important to understand the bounds of what you've uncovered. The first thing executives will ask is how do you know this is everything, that these are the only types of devices? To that end, don't just bucket devices as servers, routers, and switches. Go into the next level of detail. Executives are always going to ask for more and more detail, so make sure you're ready.

The other important part of refining your list of priorities is to understand your business processes. If I was asked about remediating critical vulnerabilities on drones, that's a very different question depending on the company I work for. At HSBC, the business context is probably a marketing team doing a video shoot, and it's maybe less important. At Amazon, that's a business-critical device delivering packages.

You have to really understand what that device is or what that system does in the context of your business—what it means to have those assets on your network. Only then can you start to prioritize.

8.1.3 Vulnerability Communication

Going from an immature vulnerability management program to orchestrating a remediation engine for your entire company is hard. It requires constant communication.

When you start out, teams are used to Windows updates and maybe some core kernel updates to certain Linux distros. They might be drowning quickly once you bring in all the data and threat feeds, so make sure they understand how to use the tools and understand the process and what the formulas actually are and mean. Teach them to filter down and take in the context.

Especially when you first set out, having a daily standup, weekly standup, or whatever frequency makes the most sense with your IT organization can help you gain traction.

You also have to communicate what you're doing and get sign-off from leadership. Your security leadership may agree to the priorities among your assets and the business context, but IT leadership has to agree to that as well.

If IT operations folks don't agree or don't appreciate the program, you'll run into resistance. At the end of the day, your metrics and risk reduction are driven by the folks on the ground, whether that's one person automating a whole new class of vulnerabilities or an army of people remediating system by system.

Your security posture is not going to improve until they understand the new system and get on board with it. Even if the leaders are on board, if you can't get the folks on the ground on board, you're not going to see an improvement in risk reduction. You need a bottom-up and top-down approach.

8.1.4 Vulnerability Remediation

Effective remediation depends on visibility into the vulnerability space and the compensating controls on your network. Let's say a remote code execution comes out on a small, obscure router. It's a high-impact vulnerability where the default credentials were left on or there was a default credential theft on the system. If you know your business interest and your context, you know whether that vulnerability is something you should worry about.

To get to that point, you'll need to understand the outside-in perspective through vulnerability feeds that show what's happening out in the wild. That window into the types of threats and scenarios can be combined with your understanding from the inside out: This is where we have the routers, these are the scenarios that could be affected. That's how you prioritize which vulns to remediate or if you have to remediate them all.

8.1.5 What Success Looks Like

Because moving to RBVM is a process, there's no point at which you're "done." But there will be moments when you see the strategy and the mindset taking hold. The day it finally clicked at HSBC, there was a product going end of life. We were on a call with our CISO and CIO and were discussing the end-of-life software and the CIO asked why it wasn't remediated.

And the response was, well, should this be remediated? It's end-of-life software, but we had extended support. The conversation quickly shifted from yes, the tool says it's critical, to whether it was actually critical in the context of our business. That conversation was being had at the executive level.

That said to me that from the top down, folks were starting to take the program seriously and understand the shift. If you have to choose between patching a critical remote code execution that's going to affect your system or removing an old piece of software that could potentially, in a worst-case scenario, have a new critical vulnerability come out that's not patched, well that's one in the hand or two in the bush.

There were plenty of metrics and data points and we had a pretty mature program in terms of KPIs, key control indicators (KCIs), and key risk indicators (KRIs) to tell us how we were performing in terms of remediation. The KPIs and KCIs spoke for themselves, and senior leadership started having conversations that were more about risk-based approaches to remediation rather than just directives to hammer the nail and get it done as quickly as possible.

It certainly took time and effort to get there and to build a vulnerability management program as robust as what HSBC now has. It was worth every conversation and every conversion to bring the teams around to the new way of doing things. Now there's data instead of guesswork, prioritization instead of randomness, and a defined process that holds risk to an acceptable level.

9

THE FUTURE OF MODERN VM

Vulnerabilities and exploits are man-made of course—but often we treat them as natural occurrences in the ether—a weather event to be dealt with rather than a weakness or a flaw in a complex system. Imagine that they really were force majeure events. Let's treat a vulnerability like a hurricane sweeping through the installations of an SQL server or Microsoft Word.

In that conception, we as a society are struggling. If the general public was warned about a hurricane on average 100 days after we saw it develop off the coast of Louisiana, we would be tripling the budget of the weather service. Cybersecurity events are of course not acts of nature, but by any standard, we're moving too slowly.

In a *Wall Street Journal* opinion piece, V.S. Subrahmanian, Director of Dartmouth College's Institute for Security, Technology, and Society, called for a National Cyber Vulnerability Early-Warning Center [1].

Subrahmanian argues that the 133-day gap between the discovery of a vulnerability and its disclosure to the public, a period designed to allow vendors to create a patch, is much too long. He notes the sluggishness of the NVD when investigating a vulnerability, and how vendors often have no reason to cooperate with the investigation or create a patch when software is no longer supported or when the vulnerability poses little risk.

The alternative, an early warning center, would more quickly discover and prioritize vulnerabilities, working like a meteorologist searches for storm systems.

In this book, we've laid out how to discover and prioritize vulnerabilities in your environment based on risk. This decision engine is designed to prioritize the vulnerabilities that already exist in your environment and adjust as new ones emerge. Forecasting the risk a vulnerability poses as soon as it's identified is a different task that requires even more data.

This is the holy grail of vulnerability management—to look at the vulnerabilities released by the NVD on any given day and make a measured, confident prediction about which ones will pose a risk to your organization. In some ways, this is already happening. Let's take a look at what the future of modern VM looks like and where the movement is headed.

9.1 STEPS TOWARD A PREDICTIVE RESPONSE TO RISK

Twenty years ago, even 10 years ago, those who built security products were focused on building the best sensors so we could capture the right data. A lot of organizations are still stuck in that mindset of "how can I detect more?" This is not incorrect. The asset coverage (percentage of assets regularly scanned) is low across all but the newest (read: most cloud-based) or most sophisticated organizations. If as a security manager you know that your organization is only scanning 10% of your assets, then more detections, better scanning, and new methods of scanning all make sense as a solution. The problem is that there's already too much noise in the queue.

This has led to a 20-year-long game of whack-a-mole—teams try to detect vulnerabilities and remediate as many as possible. We discover new vulnerabilities, manufacturers release new patches, defenders orchestrate the dance of remediation.

As we've explored throughout this book, simply showing more data to security practitioners doesn't solve any problems. In fact, it only complicates the jobs of overworked and understaffed teams that already have a long list of vulnerabilities to fix.

A vulnerability analyst will see most vulnerabilities as important—CVSS scores 28% of all vulnerabilities a 10/10—but the problem security teams need to solve is how to show IT ops which vulnerabilities are actually risky. The key here is that risk is not only a measurement of the state of an enterprise, it is also a language for communicating with the rest of the business. In finance, this "risk as language" framework is commonplace throughout the global 2000, but security has yet to catch on.

Vulnerability management is ultimately an attempt to stay ahead of attacker tactics. Whether it's scrambling to fix the latest Patch Tuesday release or chasing down Heartbleed when that vulnerability was all over the news, there's an inherent assumption that we must beat attackers to the punch, to remediate before they have the chance to exploit.

A true predictive response to risk—an early warning system—would need a national center or a collaboration between public and private organizations with a solid data backbone. It's not just people coming together to create policies and frameworks. The National Weather Service only works as well as it does at forecasting the weather because it's a great data research center. Something similar for cybersecurity would allow us to better forecast which vulnerabilities pose the greatest risk.

Of course, it's not just a data collection problem, it's also a signal-to-noise problem. Security suffers greatly from an overabundance of noise, and alerts are probably one of the loudest noises today. Often, the problem isn't that teams are missing critical alerts, it's that they also received a million other alerts that didn't mean much of anything. Both false positives and false negatives decrease efficiency.

Reducing the noise problem starts with gathering ground truth data—the data sources that tell us enough to train a model that describes which vulnerabilities are important and which ones aren't. In some areas of cybersecurity, we're starting to see improvements because we've identified some ground truth data.

That data consists of either a successful exploitation of a vulnerability or pieces of the puzzle put together. For example, you see an asset that has a vulnerability. You see an attack that matches a signature for the exploit of that vulnerability. After that signature event, you see one or more indicators of compromise on that asset. When you tie all three of those data points together, you can assume a successful exploitation of the vulnerability. Generating ground truth data in any domain is a process of defining outcomes, measuring those outcomes, and creating the automations that yield a pipeline of data describing those outcomes at scale.

Obviously, fear of owning up to compromises in their systems is a big hurdle to sharing for most businesses. On top of the reluctance to share, there's also the range of methods at work in documenting each incident, ones that don't always neatly line up across organizations. To get more and better sources of data, we need a common lexicon for how to label that data, as well as a sort of clearinghouse where organizations could share data anonymously with only as many identifying characteristics as needed to put the data in context, such as industry, revenue, number of employees, and so on [2].

If we create a central repository with a common lexicon to label data, drawing in incidents from across the government and private sector, we could take a big step toward more accurate vulnerability forecasting.

The first and perhaps most successful example of improved data collection in security is the Verizon Data Breach Investigations Report (DBIR), which, through the efforts of dozens of researchers and scientists, collected, categorized, and still studies incidents coded in the VERIS framework. Before the DBIR, there was no statistical, global view of what incidents looked like, who they affected, or how much financial havoc they caused. But data alone is rarely enough.

This book asks the reader to find outcome data and use it to create actionable models. Security is riddled with signal-to-noise problems, and successful solutions depend on the interplay between well-defined outcome measures and models that employ those measures in a tight feedback loop.

9.1.1 Passive Data Collection

If you use a web browser like Google Chrome, you're likely familiar with the periodic suggestions to update to the latest version. The reason you get that notification is because the application is aware of which version you're running. That data is known.

These lightweight sensors are gaining ground and present opportunities to passively collect data about assets used in your environment. Scanning can happen in the background more passively. From there, the decision engine could use that data to determine whether or not that version of the software has a vulnerability that's risky at a certain level of confidence.

Ultimately, to do that effectively will require a larger data set pulled from a global sensor network across the large cloud providers, like AWS, GCP, and Microsoft Azure, which collect so much data and telemetry that together, their data could support a strong prediction about the riskiness of a vulnerability.

This kind of data would help enterprises stay secure, but who fills that need is a question of policy that falls outside the scope of this book. It might be the major cloud providers, the government, security vendors, a consortium of enterprises, or even a startup founded by the reader. Sharing the data isn't enough. Threat sharing has been tried, though protocols like STIX and TAXII. The problem is that those providing the data need to get something in return, like actionable forecasts.

There are plenty of examples of data sharing in other fields. The Paris Climate Accords bring together the international community to report data on each country's progress toward meeting climate action goals. Taken in aggregate, they show whether the world is on track for meeting the agreed-upon progress. The medical community's sharing of data informs actions to slow the spread of disease and accelerate and improve treatments. Sharing of SARS-CoV-2 genome sequences aided the development of vaccines and tracking of variants during the COVID-19 pandemic. Even the cybersecurity community shares data at conferences such as FS-ISAC or HS-ISAC.

The problem is timing. Data sharing is rarely real time. In security especially, the swift and open sharing of data is critical to addressing vulnerabilities. Passive data collection could aid these efforts, working to provide a more complete and current picture of the vulnerability landscape that would inform more accurate assessments and forecasts of risk the moment it arises.

9.2 FORECASTING VULNERABILITY EXPLOITATION WITH THE EXPLOIT PREDICTION SCORING SYSTEM

As we discussed in Chapter 1, there are a number of industry standards for assessing the characteristics of a vulnerability and classifying its severity. NVD, for example, captures details and analysis around CVEs. CVSS denotes the severity of an exploit for a vulnerability with a score that reflects how damaging a vulnerability would be for an organization if exploited.

Among the limitations of these existing measures is their failure to assess a vulnerability's potential to be exploited. This is why 40% of all CVEs have a CVSS score of 7 or above. The potential is there, but how likely is exploitation to occur? That's tough to gauge from CVSS scores, which change infrequently by design. Apart from the risk-based decision engine we've built in these pages, there are few resources for determining the likelihood of exploitation for any given vulnerability.

Which is why contributors from RAND, Cyentia Institute, and Kenna Security created the Exploit Prediction Scoring System (EPSS). EPSS is an open, data-driven model for predicting which vulnerabilities are most likely to be exploited.

EPSS uses current threat information from CVEs and real-world exploit data to generate a probability between 0 and 1. The higher the score, the greater the probability that a vulnerability will be exploited. EPSS version 1 makes a prediction about how likely the vulnerability is to be exploited over the following 12 months.

Version 2 adds real-time scoring of common vulnerabilities and exposures (CVEs) as they are announced, serving as a forecast for a vulnerability's potential to be exploited. This will allow users to gain instant insight without having to gather data about a CVE elsewhere.

Using EPSS, you gain quantitative evidence of which new vulnerabilities you should focus on fixing and which ones, based on your company's risk tolerance, are lower priorities. Compared to using CVSS 9+ as a guide for remediation, EPSS can achieve the same degree of coverage with 78% less effort [3].

EPSS provides a model based on proprietary data from Fortinet, Kenna Security, Reversing Labs, and Alienvault along with the publicly sourced data and outside commercial data providers, including MITRE's CVE, NIST's National Vulnerability Database, CVSS scores, and CPE information. The model collects:

- Information that describes a vulnerability, such as descriptions, products, and vendors;
- Information about the vulnerability in the wild, including its prevalence, complexity, and severity;

9.2 FORECASTING VULNERABILITY EXPLOITATION

- Information about community reactions to the vulnerability, including social chatter, depth of discussions, exploits and tool publication;
- Ground truth of exploitation in the wild, such as exploits in malware, intrusion detection system (IDS) and intrusion protection system (IPS) alerts, or honeypot activity.

EPSS reads descriptive text for each CVE and scrapes for common multiword expressions and searches different repositories for exploit code. It creates a list of 191 tags encoded as binary features for each vulnerability. Risk scores are calculated based on 15 variables that correlate with exploitation.

Where EPSS is most useful is as a response to risk as it emerges. It's simply not feasible to patch 100% of all vulnerabilities that appear, nor would you want to spend the time and resources remediating vulnerabilities that pose no risk.

EPSS offers a quantitative way to address new threats as they emerge. It's an example of the kinds of tools and insights that governments should be focusing on as a way to combat accelerating cybersecurity attacks. In our efforts to prioritize remediation based on risk, our decision engine surfaces the riskiest vulnerabilities already in your environment. EPSS builds on that, scoring the riskiest vulnerabilities as soon as they appear and adjusting the scores as new data becomes available.

Of course, EPSS only measures the threat component of risk. Other aspects of the vulnerability, an organization's network, the asset on which the vulnerability lives, and more should all factor into a decision to remediate. Nevertheless, EPSS offers a way to upgrade national vulnerability management policies to conserve federal resources spent on vulnerability remediation, as well as to assess threats to national security.

The model can also scale to estimate the threat for multiple vulnerabilities, whether on a server or on up to an enterprise level. Organizations can determine the probability of exploitation of multiple vulnerabilities with different chances of being exploited to determine the overall threat to the asset or organization, as well as gauge its security posture over time by comparing probabilities.

The EPSS special interest group (SIG) is working to improve the maturity of data collection and analysis and is developing partnerships with data providers and establishing an infrastructure from which it can provide a publicly accessible interface for EPSS scores. The aim in the future is to gain access to even more data, especially around actual exploitations in the wild, that will help refine the predictions both in their timeliness and accuracy.

Weather forecasts use public data. They're based on radar and other instruments at every airport in the country, and the data is made publicly available to anyone. Whether you're watching the Weather Channel or using the Dark Sky app, they're all using the same data for different purposes.

Nothing like that exists in security. There are a lot of standards, but the only data that really exists is the National Vulnerability Database. However, there's a lot of forecasting data that we could gather to gain a better sense of what's happening. Imagine if instead of having weather forecasts based on data, people were simply given geographies and told to do weather forecasting, which is precisely how farmers operated before the Old Farmer's Almanac was published in 1792. That degree of blindness is not uncommon around the risk of vulnerabilities in security. We need real-time information about what's happening in the larger world to adequately prepare for the threats we face.

9.3 SUPPORT FROM INTELLIGENT AWARENESS

Michael Roytman recently helped a friend working with an Air Force startup that did radar data analysis. Looking at the radar around any U.S. city, there are thousands of planes in the sky every day. Normally, 99% of them are standard air traffic that no one should worry about. Once in a while, data emerges that one of them is taking the wrong trajectory. Maybe it's a hobby drone in the wrong place. Maybe a plane's approach is off. There are small indicators of intelligent data that can narrow your focus to what matters. Instead of looking at all 2 million planes currently in the sky, you can look at these 100, and bring that data to the people who will act on it. It's a physics problem. The same thing exists in security—organizations often distinguish between the risk a vulnerability poses on a server versus an endpoint—because that's a physics problem too. A vulnerability that requires user interaction on a server that's running a docker container is physically unlikely to be your point of intrusion.

Automation will continue to play a major role in measuring and managing risk. As we automate more and more, it's important to think about where the automation should be.

Certainly, the ability to create a situational awareness report for your entire organization is a win, even if it shows that the situation looks pretty bad. But there's a more strategic way to drive that awareness so that more vulnerabilities get fixed.

You don't need to present the entire landscape of hundreds of millions of vulnerabilities across your organization, and in fact, you might be more successful if you don't. For example, if you find 200,000 vulnerabilities across 10,000 assets, you might only present 2,000 of the vulnerabilities because you automated an intelligent analysis that homed in on the riskiest ones. Again, this is the signal-to-noise problem, and intelligent awareness can more selectively focus attention on the signal. Remember, most models don't just measure risk, they are also catalysts for action. The very act of describing a vulnerability as a "10" or a "critical" causes a downstream set of actions. Consider those while building the model, not after.

What is the probability that this vulnerability is going to be exploited? Do these exploits exist? What's the chance one gets written? Is the code already being exploited somewhere else? That problem doesn't need human input. It just needs intelligence.

If you look at the state of a system, you can see technical details, but there are also time-based inferences you can make based on past probabilities and similar systems that have already been breached. Most IT operations software defines a system as its current state, not its future state, but that will change.

This can help accelerate investigations as well. In the past, when you discovered an intrusion, analysts looked at the system and based on the particular states in the past, there might be hundreds of different ways the malware could have gotten in. It's their job to investigate and find which one—a major investment of time.

What if, instead, you could support those decisions by examining the intelligence data through a probabilistic model and narrow the potential culprits, showing that the most probable pathway was five Adobe Reader vulnerabilities. It could easily save them a week of initial investigation time. This is a near-term goal that's possible in the next couple years. As a society, we already have mental models for recommendation engines in everything from the social networks we use, to the movies we watch, to the credit cards suggested to us. In the future, these models will extend into day-to-day security operations.

The longer-term vision is to marry the data from the vulnerability management side with the data about a system's current state and build a model for decision support, not on remediation, but on investigation.

It doesn't end there. Machine learning can support any number of decisions. In this book we talked about how to support the decision of which vulnerabilities you should remediate first based on which ones are likely to cause compromise. The IT ops and detection and response sides of the house also have decisions to make. We're seeing an evolution from endpoint detection and response (EDR) to extended detection and response (XDR).

Think about what teams use those tools for. There's a set of decisions that they make and a corresponding set of machine learning models we can build to save them time. Whether it initially saves 20% or 90% matters less because it will improve over time. The key is marrying the data sets to start building those models. It's a combination of the right data sets, right place, and the gumption to build those models and deploy them. A lot of the industry has the data to do it; they just need to start.

9.4 THE RISE OF XDR

We've had niche products like network detection and response, endpoint detection and response, and so on for years. XDR approaches detection and response more holistically, encompassing all the one-off tools you need to do detection and response. Or as Gartner puts it, XDR is a "SaaS-based, vendor-specific, security threat detection and incident response tool that natively integrates multiple security products into a cohesive security operations system that unifies all licensed components" [4]. Forrester defines it similarly as "the evolution of EDR"—it "unifies security-relevant endpoint detections with telemetry from security and business tools such as network analysis and visibility (NAV), email security, identity and access management, cloud security, and more" [5]. But what these definitions are missing is that XDR will end up being more than the sum of its parts—as long as the data that's being combined is used in models.

XDR analyzes endpoints, networks, servers, clouds, SIEM, email, and more, contextualizing attacker behavior to drive meaningful action [6].

Most SOC teams are probably already dipping a toe in XDR, according to VMWare's Rick McElroy [7]. They're logging Windows events, have data sources on the network they're putting into a security information and event management (SIEM). The push is to build everything into the infrastructure itself instead of just bolting it on.

Right now we're building a lot of instrumentation in security, and it's how we use that data that ultimately makes our systems more or less secure. It would be great if a security engineer didn't have to cobble together 14 different tools and somehow translate that output to the scripting API call. They should be able to simply issue commands like in Python or Ruby.

According to the Enterprise Strategy Group, 31% of security professionals say they spend their time responding to emergencies, 29% acknowledge security monitoring blind spots, and 23% say it's difficult to correlate security alerts from different tools [8]. Perhaps it's no surprise that the same survey found 70% "could foresee creating an XDR budget within the next 12 months." Some 23% are already working on an XDR project.

The biggest issue might be data. Some 38% of organizations had trouble filtering out noisy alerts, 37% had trouble accommodating growing security telemetry volumes, and 34% struggled to build an effective data pipeline for stream processing.

XDR addresses alert fatigue and the ever-expanding arsenal of tools at the fingertips of security professionals.

The good news around XDR is that we've been pushing for a decade plus for a data-driven approach to security. Twenty years ago, we all complained we didn't have the data to make decisions. Now everyone's overwhelmed with data, and XDR is a way to make sense of it, allowing us to become more quant-like in our decisions around security and remediation and where we invest and don't invest. It's still very early for XDR and, as always, the devil is in the details, which in this case, means in implementation. But the consolidation of tools allows for better visibility and better decision-making based on more comprehensive data.

9.5 THE OTHER SIDE OF THE COIN: REMEDIATION

Every year we see more vulnerabilities. The rate of exploitation in the wild, however, is pretty close to static. What that means is that the haystack is getting much bigger but the number of needles you have to find is roughly the same. Figuring out what to fix will remain a problem for most organizations for the foreseeable future. Assessment and prioritization of vulnerabilities certainly gets the lion's share of attention and will continue to, since both show no signs of being any less important any time soon.

The other piece of this, which is already starting to shift, is the actual remediation. Once you assess and prioritize, how do you actually fix the vulnerabilities? We've started to detect a notable difference in how different platforms are remediated and how quickly, depending on how much automation is at play. Prioritization will continue to matter, even as automation allows organizations to remediate a greater number of vulnerabilities. Year after year, the attack surface will only grow larger. Automation will help us keep up.

In a study of vulnerability data across more than 9 million active assets in 450 organizations, we found that 85% of the asset mix was Windows-based systems. Windows platforms typically have 119 vulnerabilities detected in any month—more than four times more than the next highest, MacOS. However, Microsoft automates and pushes patches for known vulnerabilities, which speeds up the remediation process. The half-life of vulnerabilities on Windows systems is 36 days, while for network appliances, it's 369 days. Vendors play a major role in the remediation velocity on their platforms, as the survival curve in Figure 9.1 shows.

Microsoft's remediation velocity is even more impressive when you compare native Microsoft flaws to those affecting the many nonnative software on assets running a Windows OS. Figure 9.2 shows that in the first month, 68% of Microsoft vulnerabilities are remediated, while the same is true for only 30% of vulnerabilities on nonnative software living on the same assets.

Microsoft and Google have leaned in heavily on auto update features for remediation, and they've already been so successful that we suspect others will start to follow. Automation will be key both

9.5 THE OTHER SIDE OF THE COIN: REMEDIATION

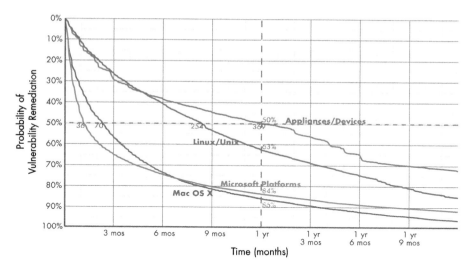

Figure 9.1 Survival analysis curves for major asset categories. (© 2020 Kenna Security/Cyentia Institute. Reprinted with permission [9].)

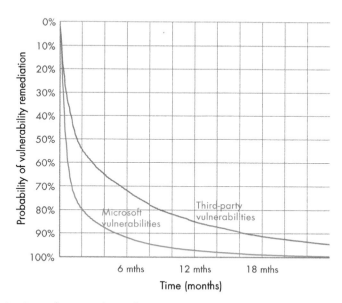

Figure 9.2 Remediation velocity for native vs nonnative vulnerabilities in Windows 10. (© 2020 Kenna Security/Cyentia Institute. Reprinted with permission [9].)

in prioritizing which vulnerabilities to fix, as well as for platforms remediating known vulnerabilities to ensure their products are secure.

Over time, we'll start to see even more immutable infrastructure as it relates to remediation. Organizations won't be patching as much as replacing.

We're already seeing this on the cloud side with infrastructure as code. When you're updating an existing system, there are so many complexities and so much that can break. The concept of immutable infrastructure, where you turn up a new container that has all of the updated services within it, and turn down the old one, makes the process a lot cleaner. Automation can handle that relatively quickly and smoothly. Technologies like Terraform are already taking advantage of this ability. This way organizations can stand up a new and updated environment in minutes or hours instead of weeks or months.

The advent of cloud—specifically infrastructure as a service—enables that. The growth of software as a service (SaaS) means there's a lot more concentrated responsibility for updates. If you're a customer of a SaaS application like Salesforce, for example, it's Salesforce's job to update everyone. It's Google's job to update its applications. They can make one update to a centralized platform and all the customers using it benefit from that as opposed to all of the individual customers having to make their own updates.

9.6 THE WICKED PROBLEM OF SECURITY ADVANCES

As we learned in Chapter 2 when we talked about game theory and wicked problems, everything this book details is a first step. For every move we make in security, our adversaries will make one in response. Based on that response, we can make another of our own. Maybe it will be investing more time and money into data. Maybe we need to invest more time and money into the modeling or into building more secure software.

Today we know that one failure point is the quality of available data. Once we fix that, the system will change again. It's a wicked problem. We know the problem today is vulnerability management. It hasn't gotten to the modern state just yet. Once it does, new problems will arise. That's how security works. First, we need to solve this one. This book is a story about how the authors brought together different domains and expertise in an attempt to solve the vulnerability management problem. It is our hope that this book endures as a practical guide about how to solve new those new problems. We leave the rest to you.

References

[1] Subrahmanian, V. S., "Cybersecurity Needs a New Alert System," *The Wall Street Journal*, March 29, 2021, https://www.wsj.com/articles/cybersecurity-needs-a-new-alert-system-11617039278.

[2] Barrachin, M., and A. Pipikaite, "We Need a Global Standard for Reporting Cyber Attacks," *Harvard Business Review*, November 6, 2019, https://hbr.org/2019/11/we-need-a-global-standard-for-reporting-cyber-attacks.

[3] Forum of Incident Response and Security Teams (FIRST), "Exploit Prediction Scoring System (EPSS)," https://www.first.org/epss/model.

[4] Trellix, "What Is Extended Detection and Response (XDR)?" https://www.trellix.com/en-us/security-awareness/endpoint/what-is-xdr.html.

[5] Cortex, "What is XDR?" https://www.paloaltonetworks.com/cyberpedia/what-is-xdr.

[6] Drolet, M., "EDR, XDR and MDR: Understanding the Differences Behind the Acronyms," *Forbes*, April 15, 2021, https://www.forbes.com/sites/forbestechcouncil/2021/04/15/edr-xdr-and-mdr-understanding-the-differences-behind-the-acronyms/?sh=343efc5f49e2.

[7] Personal communication, https://podcast.kennaresearch.com/public/46/Security-Science-80c4443c/fa09746a/.

[8] Oltsik, J., "The Realities of Extended Detection and Response (XDR) Technology," *Dark Reading*, February 24, 2021, https://www.darkreading.com/vulnerabilities-threats/the-realities-of-extended-detection-and-response-xdr-technology.

[9] Kenna Security and The Cyentia Institute, *Prioritization to Prediction Volume 5: In Search of Assets at Risk*, 2020. https://learn-cloudsecurity.cisco.com/kenna-resources/kenna/prioritization-to-prediction-volume-5.

GLOSSARY

Area under the survival curve (AUC) Representation of "live" (open) vulnerabilities. A lower AUC means higher velocity.

Asset Hardware or software connected to a network.

Asset-centric A model of vulnerability remediation focused on the business criticality, value of, and exposure of an asset, with an approach of gradual risk reduction.

Bugs Flaws that cause unintended effects in an asset.

Closed-source intelligence A type of threat intelligence feed based on dark web forums, intelligence agencies and law enforcement, and human intelligence.

Completeness The most current and accurate knowledge of as many assets as possible in your environment.

Coverage Measure of the completeness of vulnerability remediation, or the percentage of exploited or high-risk vulnerabilities that have been fixed.

Edge intelligence A type of threat intelligence feed based on host activity at the edge of a network.

Efficiency Measure of the precision of remediation, such as the percentage of all remediated vulnerabilities that are actually high risk.

Exploit Code that compromises a vulnerability.

Exploitation An event where an exploit is used to take advantage of a vulnerability.

Incremental probabilities A way to communicate probabilities and scores by adding in one variable at a time.

Internal intelligence A type of threat intelligence feed based on an organization's own assets and behavior.

k-means clustering A common strategy of malware classification that groups data points into k clusters based on common features, with the mean of the data representing the center of each cluster, known as the centroid vector.

Machine learning (ML) A classification of artificial intelligence in which algorithms automatically improve over time by observing patterns in data and applying those patterns to subsequent actions.

Mean time to detection (MTTD) The average amount of time it takes to detect vulnerabilities.

Mean time to remediation (MTTR) The average amount of time it takes to close vulnerabilities.

Modern vulnerability management An orderly, systematic, and data-driven approach to enterprise vulnerability management.

Natural language processing (NLP) Using text data to create new variables in supervised learning ML applications.

Network intelligence A type of threat intelligence feed based on traffic at an organization's network boundary and on external networks.

Neural nets A multistage method of generating coefficients to make a prediction based on the interrelated elements that are most indicative of the outcome.

Objective functions Metrics used to construct machine learning models.

Open-source intelligence A type of threat intelligence feed based on publicly available sources.

Principal component analysis (PCA) A method of data analysis that improves efficiency and the creation of predictive models by reducing the dimensionality of multivariate datasets so they are easier to analyze and interpret.

Random forests A collection of decision trees randomly created from a selection of features in the larger dataset.

Regressions Regression algorithms take the outcome measure, add several variables, and find a coefficient for each one. The variables that correlate with the outcome measure will allow the algorithm to then predict unknown outcomes based on known parameters.

Remediation capacity The number of vulnerabilities that can be remediated in a given time frame and calculates the net gain or loss.

Remediation velocity Measure of the speed and progress of remediation.

Risk According to ISO Guide 73, risk is the "combination of the probability of an event and its consequence."

Risk-based vulnerability management (RBVM) A cybersecurity strategy in which you prioritize remediation of vulnerabilities according to the risks they pose to your organization.

Risk meter A way to communicate progress in reducing the risk of vulnerabilities by department, asset group, or other category.

Risk scores A number on a scale of 1 to 100, with 100 representing the highest risk, that's assigned to each vulnerability in an organization to represent the relative risk the vulnerability poses and help security teams prioritize and manage that vulnerability.

Threat The potential for a specific actor to exploit, or take advantage of, a vulnerability.

Tolerance An amount of risk that an individual or organization deems acceptable.

Threat-centric A model of vulnerability remediation focused on vulnerabilities actively targeted by malware, ransomware, exploit kits, and threat actors in the wild, with an approach geared toward imminent threat elimination.

Top influencers A way to communicate probabilities and scores by determining the major influencers for a vulnerability rating.

Top influencers by proportion A way to communicate probabilities and scores in which each weight is treated as independent but as part of the larger proportion to estimate its overall influence.

Vulnerabilities The flaws or weaknesses in assets that could result in a security breach or event.

Vulnerability assessment The process of scanning and analyzing assets for existing vulnerabilities.

Vulnerability-centric A model of vulnerability remediation focused on the criticality of a vulnerability, with an approach of gradual risk reduction.

Vulnerability half-life Time required to close exactly 50% of open vulnerabilities.

Vulnerability management Efforts at mitigating relevant vulnerabilities.

ABOUT THE AUTHORS

Michael Roytman is the chief data scientist of Kenna Security, now a part of Cisco. His work focuses on cybersecurity data science and Bayesian algorithms, and he served on the boards for the Society of Information Risk Analysts, Cryptomove, and Social Capital. He was the cofounder and executive chair of Dharma Platform (acquired, BAO Systems), for which he landed on the 2017 "Forbes 30 under 30" list. He currently serves on Forbes Technology Council. He holds an MS in operations research from Georgia Tech, and has recently turned his home roasting operation into a Chicago South Side cafe, Sputnik Coffee.

Ed Bellis is the cofounder and CTO of Kenna Security, now a part of Cisco. He is a security industry veteran and expert. Known in security circles as the father of risk-based vulnerability management, Bellis founded Kenna Security to deliver a data-driven, risk-based approach to vulnerability management and help IT teams prioritize their actions to meaningfully reduce cybersecurity risk. Ed is the former CISO of Orbitz and the former vice president, Corporate Information Security at Bank of America before that. Ed is a frequent speaker at industry conferences including RSA and Black Hat, and a cybersecurity contributor to Forbes, Dark Reading, SC Magazine, and other publications.

INDEX

A

Access control lists (ACLs), 22
Activations, 84
Akaike information criterion (AIC), 67
Alerts, speed of, 147–49
Analytical models, 34
Apache Struts vulnerability, 2–3
Area under the survival curve (AUC), 10, 125
Asset-centric model, 22
Assets
    data snapshot showing, 32
    decisions and1, 151–52
    discovery and categorization, 62–64
    inventory, 55, 56
    management assessment, 143–44
    starting with, 141–42
Attackers
    defenders interaction, 33
    finding vulnerabilities, 36
    game theory and, 33
    incentives and, 34
    vulnerability management and, 189
Authenticated scans, 55, 56
Automation, 134, 160
Average remediation rates, 114–18

B

Batching, 153–54
BlueKeep, 126
Bottomry, 16
Building a system
    about, 141

aggregation and correlation in, 149
asset management assessment and, 143–44
assets and, 141–42
considerations before, 141–46
database architecture, 150–54
organizational direction and, 144–45
premise versus cloud and, 146–47
processing considerations, 147–50
role-based access controls, 156–57
scale and, 142
search capabilities, 154–56
SOC volume, 149–50
speed of decisions and alerts and, 147–49
tools as constraints in, 145–46
users and, 142
vulnerability density and, 142

C

Calibration plot, 69
Capacity
    about, 123–24, 162
    concept, 32–33
    cost of, 131
    determination, 130
    limits, 177–78
    power law of, 132
    prioritization and, 131
    remediation, 114–15, 117, 130
    team structure and, 132
Capping, 35

Change management, 174–75
Chatter and Exploit database, 60
Closed-source intelligence, 62
Cloud, the
  on premise versus, 146–47
  vulnerability debt and, 134–35
Clustering
  about, 41–43
  data, 41–43
  k-means, 43–44
  for malware classification, 43–44
Common Platform Enumeration (CPE), 7–8, 58, 84, 85
Common vulnerabilities and exposures (CVEs)
  about, 5–7
  age of, 81, 84, 85
  as application security findings, 59
  comparison of, 9
  enumeration, 59
  life cycle, 60–61
  patching process and, 125–26
  predicted probability between, 69
  "really bad," prioritizing, 9–10
  "really critical," 120
  references, 80
  start year, 84–86
  velocity, 126
  year of, 81
Common Vulnerability Scoring System (CVSS)
  about, 2, 7
  hyperparameter, 86, 87
  information, 80
  for remediation, 10, 110
  See also CVSS scores
Common Weakness Enumeration (CWE)
  about, 7
  example, 57–58
  hyperparameter, 86, 88
  mappings, 60
  nested hierarchy, 59–60
  types of, 57
Communication, 161, 183–84
Compliance, 161
Confidentiality, integrity, and availability (CIA), 80
Configuration management database (CMDB)
  about, 62–63
  coordination with, 64
  functions, 63–64

Coverage
  about, 9, 119–20, 161
  efficiency tradeoffs, 120–21
  plotting, 121, 123, 124
  in real world, 121–23
  scanning, 109
CVE-2021-21220, 50–55
CVE Numbering Authority (CNA), 5–7
CVSS 7+, 120
CVSS 10+, 122
CVSS scores
  about, 20–21
  change and, 121
  data for creation of, 80
  remote code execution vulnerabilities and, 45
  temporal, 80, 86, 88
  See also Common Vulnerability Scoring System (CVSS)
Cybersecurity
  attackers and defenders in, 32
  challenges facing, 18–19
  exploits, 25–26
  fundamental terminology, 4–8
  machine learning for, 38–45
  risk management and, 15–24
  scale of, 18–20

D

Data
  about, 48–49
  asset discovery and categorization, 62–64
  cleansing, 65
  collection, passive, 190–91
  definitions versus instances, 50–55
  formats, 49
  loading, 65
  preparation of, 72, 79–82
  quality, 48
  sources, logistic regression model, 66
  strategies of approach, 49
  threat intel sources, 60–62
  validating and synthesizing, 49, 64–65
  vulnerability assessment, 55–60
Database architecture
  about, 150–51
  assets and decisions and, 151–52
  illustrated, 151
  real-time risk management and, 152–54
Data Breach Investigations Report (DBIR), 190

Data clustering, 41–43
Decision engine
   about, 47–48
   data, 48–65
   logistic regression model, 65–79
   neural network design, 79–101
Decisions
   assets and, 151–52
   speed of, 147–49
   support systems, 26
Descriptive statistics, 29
Distributed decision-making, 175–77
Dynamic application security testing (DAST), 58, 59

E

Edge intelligence, 62
Efficiency
   about, 9, 119–20, 161
   concept, 32–33
   coverage tradeoffs, 120–21
   improving, 161
   plotting, 121, 123, 124
   in real world, 121–23
   velocity and, 129–30
Empirical models, 34
Empirical normalization, 97
Endpoint detection and response (EDR), 196
Epidemiology, 30–31
Equifax portal incident, 1–3
Exception handling, 179
Exchange servers, 127–28
Executive buy-in, 166–67
Exploitation
   about, 4
   forecasting, 191–94
   variables, 80, 86, 89, 95
Exploit Prediction Scoring System (EPSS)
   about, 191–92
   descriptive text, 193
   forecasting with, 191–94
   measurement, 193
   model, 192–93
   special interest group (SIG), 193–94
   threat information, 192
Exploits
   about, 4
   as adversary, 25
   cybersecurity, 25–26
   evolution of, 26
Exposures, common, 5–7

Extended detection and response (XDR), 196–97
Extensible Markup Language (XML), 65
Extract, transform, and load (ETL) process, 64–65

F

Far-field radiation magnitude, 60–62
Feature engineering, 66–68
Features, interpretation of, 68
Firewall rules, 22

G

Game theory, 32–34
Gamification, 167–68
Geer, Dan, 18–19, 23
Glossary, 203–6
Good metrics, 105–11

H

Hidden layers, 93, 94
Hidden units, 93, 94
Hyperparameter exploration/evaluation
   about, 84
   CPE, 84, 85
   CVE age, 84, 85
   CVE start year, 84–86
   CVSS, 86
   CVSS temporal, 86, 88
   CVSS variables, 86, 87
   CWE, 86, 88
   exploit variable, 86, 89
   hidden layers, 93, 94
   hidden units, 93, 94
   product variables, 89, 90
   reference variables, 89, 90
   tags, 89, 91
   vendor information, 89, 91
   volatility variables, 89, 92
   vulnerability counts, 89, 92

I

Implementation in production
   about, 72
   application of model, 72–76
   binary variables, 73
   communication of results, 77–79
   converting log odds to probability, 76–77
   data preparation, 72
   model coefficients, 74

Implementation in production (continued)
  See also Logistic regression model
Incremental probabilities, 77–78
Inferential statistics, 29
Influencers, top, 78–79
Instances, 55
Intelligent awareness, 194–96
Interdependencies, 27–28, 30
Internal intelligence, 62

K

Key control indicators (KCIs), 185
Key performance indicators (KPIs), 161, 185
Key risk indicators (KRIs), 185
K-means clustering, 43–44

L

Latin hypercube sampling, 84
Logistic regression model
  attributes, 65–66
  binary variables, 73
  building, 65–79
  calculations, 75, 76
  coefficients, 74
  data sources, 66
  feature engineering, 66–68
  implementation in production, 72–79
  interpretation of features, 68
  objectives, 65
  performance testing, 69–72
Log odds, converting to probability, 76–77

M

Machine learning (ML)
  about, 38–39
  algorithm types, 39
  data clustering and, 41–43
  neural nets and, 40, 43, 79–101
  NLP and, 44–45
  PCA and, 38–39
  random forests and, 40, 42
  regressions and, 40, 41, 42
  supervised models, 39–40
  unsupervised models, 40–45
Malware, clustering for classification, 43–44
Markov chains, 35–36, 37
Mathematical modeling
  about, 26
  game theory, 32–34
  interdependencies and, 27–28

NP problem and, 28
OODA loops, 37–38
scale, 27–29
statistics, 29–31
stochastic processes, 34–37
Mathematical scale, 27–28
Maturity process, 159
Mean monthly close rate (MMCR), 11, 130
Mean time to detection (MTTD), 10, 112
Mean time to exploitation (MTTE), 112
Mean time to incident discovery, 108–9
Mean time to remediation (MTTR), 10, 112
Mean-time-tos, 111–12
Media-based vulnerabilities, 178–79
Meltdown, 125–26
Metrics
  automatic computation, 108
  bounded, 106
  capacity, 32–33, 114–15, 123–32, 162
  context-specific, 107–8
  coverage, 9, 109, 119–23, 161
  efficiency, 9, 32–33, 119–23, 129–30, 161
  evaluation of, 108–11
  "good," 105–11
  new, reporting, 167
  objective, 106
  reliable, 107
  remediation, 111–18
  scaling metrically, 106
  Type I and Type II, 107
  valid, 107
  velocity, 112–14, 123–32, 162
Model performance testing
  about, 69
  calibration plot, 69
  simplicity versus performance and, 70–72
  trade-off visualization, 71
  variables, 71
Modern vulnerability management (MVM), 55
Monte Carlo simulations, 36

N

National Vulnerability Database (NVD)
  about, 7
  CVE definitions, 50
  CWE and CPE and, 58
Natural language processing (NLP) and, 44–45
Network intelligence, 62

Network scans, 55, 56
Network topology, 33
Neural nets
  about, 40, 43, 79
  activations, 83
  architecture, 82–84
  data preparation, 79–82
  designing, 79–101
  future work, 100
  layers, 83
  model development, 82–84
  regularization, 83
  scoring, 95–100
  size, 83
  smoothing, 83–84
  stopping, 84
  wide, narrow, or constant, 83
Normalization, 97, 98, 99

O

Observe-orient-decide-act (OODA) loops, 37–38
Open-source intelligence, 62
Operations research, birth of, 16–18

P

Passive data collection, 190–91
Patching process, 125–26
Payment Card Industry Data Security Standard (PCI DSS), 20
Performance
  importance of, 118–19
  risk versus, 104–5
  testing, 69–72
Performance measurement
  about, 103–4
  coverage and efficiency, 119–23
  "good" metrics and, 105–11
  remediation metrics, 111–18
  remediation SLAs, 135–38
  use of, 118–19
  velocity and capacity, 123–32
  vulnerability debt, 132–35
Personally identifiable information (PII), 1
Power law, 132, 138
Predicted probabilities, 69, 70
Principal component analysis (PCA), 38–39
Prioritization
  automation, 160
  capacity and, 131
  data-driven, 3

failures, 3
move to the cloud and, 134
risk scores and, 162–64
vulnerability, 22
vulnerability assessment and, 182–83
Probabilities
  converting log odds to, 76–77
  incremental, 77–78
  predicted, 69, 70
Process, importance of, 179
Product variables, 89, 90
Proofs of concept (POCs), 80, 101
P versus NP, 27–28

R

Random forests, 40, 42
Random walk, 34, 35, 37
Real-time risk management
  about, 152–53
  batching and, 153–54
  data point collection and, 153
  vulnerability forecasts and, 153
Real-world examples
  about, 181–82
  success and, 184–85
  vulnerability assessment and prioritization, 182–83
  vulnerability communication, 183–84
  vulnerability discovery, 182
  vulnerability remediation, 184
Rectified linear unit (ReLU) activation, 83
References, 80
Reference variables, 89, 90
Regressions, 40, 41, 42, 65
Regularization, 83
Remediation
  capacity limits, 177–78
  coverage and efficiency and, 119–23
  deadlines, by priority level, 137
  effective, 184
  Microsoft and Google, 198–200
  models, 22–23
  SLAs, 135–38
  traditional strategies, 120
  velocity and capacity, 123–32
Remediation metrics
  average rates, 114–18
  capacity, 114–15, 117
  mean-time-tos, 111–12
R values, 114–18
  volume and velocity, 112–14
Remote code execution vulnerabilities, 45

Reporting
    improving, 161
    new metrics, 167
    as a service, 155–56
Rescoring, 149
Return on investment (ROI), 31
Risk
    about, 4–5, 104
    data snapshot showing, 32
    decision engine for forecasting, 47–101
    defining with data science, 15–24
    driving down, 164–68
    levels, 4
    meter, 162
    performance versus, 104–5
    predictive response to, 188–91
    processing, 5
    scoring, 47
Risk-based vulnerability management (RBVM)
    about, 11–12
    common goals and risk measurements, 160–62
    culture and, 163–64
    data snapshot, 32
    maturity process and, 159
    in modern vulnerability management, 12
    origins of, 20–24
    precursors to, 21
    shift to, 160–64
Risk hunting, 155
Risk management
    approaches, 16
    history and challenges, 15–24
    real-time, 152–54
Risk scores
    good, building, 47
    intelligence in generating, 11
    prioritization, 162–64
    scale, 162
    shifting to, 171
Role-based access controls, 156–57
Router rules, 22
R values, 114–18

S

Sampling, Latin hypercube, 84
Scaling
    in building a system, 142
    metrically, 106
    score, 95–96
    volume, 96
Scanning, 63, 109
Scores, combining, 96–97
Score scaling, 95–96
Scoring
    about, 95
    combining scores and, 96
    comparison to existing model, 97–100
    score scaling and, 95–96
    volume scaling and, 96
Search capabilities
    about, 154
    reporting as a service and, 155–56
    risk hunting versus threat hunting and, 155
Security advances, 200–201
Security Content Automation Protocol (SCAP), 22
Security information and event management (SIEM), 197
Security metrics, 8–13
Self-service
    change management and, 174–75
    distributed decision-making and, 175–77
    maturity, signs of, 177
    shifting to, 173–77
Service level agreements (SLAs)
    about, 135
    adherence, 168–73
    contents of, 172–73
    getting up to speed with, 171–72
    importance of, 173
    priority-driven, 137
    remediation, 135–38
    risk-based, 150
    starting with, 136
    status of, 136
    when to implement or revise, 170–72
Smoothing, 83–84, 93–95
SOC volume, 149–50
Static application security testing (SAST), 58, 59
Statistical Research Group (SRG), 16
Statistics
    descriptive, 29
    epidemiology and, 30–31
    inferential, 29
    interdependencies and, 30
    ROI and, 31
Steady-state workflow, 177–79
Stochastic processes

approaches to, 34
Markov chains, 35–36, 37
modeling uncertainties, 37
random walk, 34, 35, 37
Stopping, 84
Structured query language (SQL), 59
Supervised models, 39–40

T

Tags, 80, 89, 91
Teams
    aligning with goals, 165–66
    importance of, 179
    reducing friction between, 161
    size of, 174
    structure of, 131–32
Threat-centric model, 22
Threat feeds
    about, 60
    categories, 62
    use example, 61–62
Threat hunting, 155
Threats, far-field radiation magnitude, 60–62
Top influencers, 78–79
Training, validation, and testing, 81

U

Uncertainties, modeling, 37
Unsupervised models, 40–45

V

Variables
    binary, of model, 73
    breaking into classes, 79
    CVSS, 86, 87
    CWE, 86, 88
    distribution of predictions for, 95
    exploitation, 80, 86, 89, 95
    influencing, 78–79
    model performance testing, 71
    nonexploited, 95
    number of flips across, 82
    product, 89, 90
    reference, 89, 90
    tag, 89, 91
    volatility, 89, 92
Velocity
    about, 123, 162
    CVEs, 126
    efficiency and, 129–30
    factors impacting, 124–25
    remediation, for product vendors, 126
    team structure and, 131
    variation, 128–29
Vendor information, 89, 91
Volatility, 81, 89, 92
Volume
    remediation, 112–14
    scaling, 96
    SOC, 149–50
Vulnerabilities
    Apache Struts, 2–3
    common, 5–7
    data snapshot showing, 32
    defined, 4
    definition, 50
    discovery and categorization, 182
    forecasting, 47–48
    growth in, keeping up with, 160
    high-risk versus low-risk, 169–70
    instances, 55
    landscape, state of, 1
    management, 3
    media-based, 178–79
    numbers of, 19–20
    open and closed, average, 116, 117
    patches for, 21
    prioritization, 22
    problem of fixing, 28–29
    reaction to, 30–31
    remote code execution, 45
    superhot, 150
Vulnerability assessment
    about, 29–30
    asset inventory and, 55, 56
    authenticated scans and, 55, 56
    blind spots, 56–57
    data, 55–60
    methods, 55
    network scans and, 55, 56
    prioritization and, 182–83
Vulnerability-centric model, 22
Vulnerability counts, 80, 89, 92
Vulnerability debt
    about, 132–33
    illustrated, 133
    move to the cloud and, 134–35
    paying down, 135
Vulnerability forecasts, 153
Vulnerability half-life, 10
Vulnerability lifecycle milestones, 127
Vulnerability management
    attackers and, 189

Vulnerability management (continued)
   program development, 12
   risk-based (RBVM), 11–12, 20–24
Vulnerability scoring model
   attributes, 65–66
   binary variables, 73
   building, 65–79
   calculations, 75, 76
   coefficients, 74
   data sources, 66
   feature engineering, 66–68
   implementation in production, 72–79
   interpretation of features, 68
   objectives, 65
   performance testing, 69–72
Vulnerability survival analysis curves, 114, 115, 125

W

Wald, Abraham, 17–18

Recent Titles in the Artech House Computer Security Series

Rolf Oppliger, Series Editor

Bluetooth Security, Christian Gehrmann, Joakim Persson, and Ben Smeets

Computer Forensics and Privacy, Michael A. Caloyannides

Computer and Intrusion Forensics, George Mohay, et al.

Contemporary Cryptography, Second Edition, Rolf Oppliger

Cryptography 101: From Theory to Practice, Rolf Oppliger

Cryptography for Security and Privacy in Cloud Computing, Stefan Rass and Daniel Slamanig

Defense and Detection Strategies Against Internet Worms, Jose Nazario

Demystifying the IPsec Puzzle, Sheila Frankel

Developing Secure Distributed Systems with CORBA, Ulrich Lang and Rudolf Schreiner

Electric Payment Systems for E-Commerce, Second Edition, Donal O'Mahony, Michael Peirce, and Hitesh Tewari

Engineering Safe and Secure Software Systems, C. Warren Axelrod

Evaluating Agile Software Development: Methods for Your Organization, Alan S. Koch

Implementing Electronic Card Payment Systems, Cristian Radu

Implementing the ISO/IEC 27001 Information Security Management System Standard, Edward Humphreys

Implementing Security for ATM Networks, Thomas Tarman and Edward Witzke

Information Hiding, Stefan Katzenbeisser and Fabien Petitcolas, editors

Internet and Intranet Security, Second Edition, Rolf Oppliger

Introduction to Identity-Based Encryption, Luther Martin

Java Card for E-Payment Applications, Vesna Hassler, Martin Manninger, Mikail Gordeev, and Christoph Müller

Lifecycle IoT Security for Engineers, Kaustubh Dhondge

Modern Vulnerability Management: Predictive Cybersecurity, Michael Roytman and Ed Bellis

Multicast and Group Security, Thomas Hardjono and Lakshminath R. Dondeti

Non-repudiation in Electronic Commerce, Jianying Zhou

Outsourcing Information Security, C. Warren Axelrod

The Penetration Tester's Guide to Web Applications, Serge Borso

Privacy Protection and Computer Forensics, Second Edition, Michael A. Caloyannides

Role-Based Access Control, Second Edition, David F. Ferraiolo, D. Richard Kuhn, and Ramaswamy Chandramouli

Secure Messaging with PGP and S/MIME, Rolf Oppliger

Securing Information and Communications Systems: Principles, Technologies and Applications, Javier Lopez, Steven Furnell, Sokratis Katsikas, and Ahmed Patel

Security Fundamentals for E-Commerce, Vesna Hassler

Security Technologies for the World Wide Web, Second Edition, Rolf Oppliger

Techniques and Applications of Digital Watermarking and Content Protection, Michael Arnold, Martin Schmucker, and Stephen D. Wolthusen

User's Guide to Cryptography and Standards, Alexander W. Dent and Chris J. Mitchell

For further information on these and other Artech House titles, including previously considered out-of-print books now available through our In-Print-Forever® (IPF®) program, contact:

Artech House
685 Canton Street
Norwood, MA 02062
Phone: 781-769-9750
Fax: 781-769-6334
e-mail: artech@artechhouse.com

Artech House
16 Sussex Street
London SW1V HRW UK
Phone: +44 (0)20 7596-8750
Fax: +44 (0)20 7630-0166
e-mail: artech-uk@artechhouse.com

Find us on the World Wide Web at: www.artechhouse.com